Cool Math

Scenarios and Strategies

For elementary to middle school students and teachers

Ron Dai

ISBN-13: 978-1981218608
ISBN-10: 1981218602

Acknowledgement

In this book, some pictures are adapted from online public image resources. Regarding copyright concerns, please contact the author.

About the Author

Ron Dai is a mathematics and computer science class instructor at NWCS, teaching since 2004. He has worked as a coach in the school math Olympic program for more than 10 years, and received a 2007 Junior High School Mathematics Teaching Edyth May Sliffe Award. His email contact is rong_dai@hotmail.com.

Table of Contents

Introduction

Learning math is fun. However, without the proper guidance, learning math can also be painful. This book is written to provide an enriched resource to students who yearn for an in-depth mathematical education. In addition, math coaches can use its organized scenario-based approach to teach students how to learn math more intelligently and effectively.

To Students

If you are looking for a book that illuminates the boundless applications of elementary mathematics in your daily life, or are currently preparing for math competitions and in need of a comprehensive tool to enhance your familiarity with the challenging world of math, you have chosen the right book. Consisting of hundreds of selected problems from recent AMC 8 and MathIsCool tests, this book is organized into scenario-based categories that will deliver concepts to you sequentially. The problems are all methodically designed. Each chapter usually begins with a simple story that leads into explanations of basic concepts through numerous examples. Please make sure you understand the idea behind each solution, before you proceed to tackle practice problems.

To Teachers

Having the right resources is indispensable for a math teacher. This book provides plenty of problem solving examples in well-organized sections with 200+ follow-up exercises located at each chapter's conclusion. For your reference, their answer keys have been added at the end of the book. Depending on the curriculum you have designed for your class, you may choose to use this book as a supplemental resource for teaching content on specific topics. I hope this book becomes a useful extension to your teaching arsenal.

The American Mathematics Competition (AMC), organized by the Mathematical Association of America (MAA), is a series of annual math competition events held nationwide. The AMC 8 exam is designed for students in 8th grade and below. See details on http://www.maa.org, the official website of the MAA.

"MathIsCool" is a math competition exclusive to the Pacific Northwest region. Organized by a group of highly-experienced math-teaching experts in Washington state, the event has been very successful for 20 years and is even expanding its influence to neighboring states. You may refer to http://academicsarecool.com/history.php for details.

Clock

[Basics] time conversion, circle, angle, ratio, fraction, Least Common Multiple

Clocks are among the most important tools in our daily lives. There are two types of clocks: a traditional analog clock, and a "modern" clock with a digital display. The analog clock is usually in a round shape. The reason is simple. The clock is supposed to be "running" forever without stopping or pausing. The natural phenomenon that the earth circulates around the sun reminds us that a circular shape is the best way to design a clock. There are two "needles", a minute hand and an hour hand, on a clock. Sometimes there is a third hand, the "second" hand. If you keep watching how they move on a clock, you can find the minute hand is chasing the hour hand all the time. You may associate it with two racers running on a round racing track, where the "minute racer" is much faster than the "hour racer". During a 24-hour period, the minute hand overlaps the hour hand many times. Unless you stop the minute hand, like what the hero did in a thrilling and classic spy movie - "The 39 Steps", it keeps moving until the battery goes dead, or the gears run out of grease if it is a mechanical clock. A funny thing is, even if a clock stops working, it may still be useful at two moments during a day. This is why some say that, "A clock never showing the right time might be preferable to the one showing the right time only twice a day."

[Question] Do you know how many times the minute hand meets the hour hand from the early morning at 12:00 AM (exclusively) to noon at 12:00 PM?

[Answer] A simple and direct approach is to list what happens during each hour period. Obviously, they meet each other 11 times during the 12-hour period. Note the starting point when they meet at 12:00 AM is the same time moment as they meet at 12:00 PM.

12:00 AM – 01:00 AM	No overlaps after 12:00 AM
01:00 AM – 02:00 AM	Overlaps between 01:05 and 01:10
02:00 AM – 03:00 AM	Overlaps between 02:10 and 02:15
03:00 AM – 04:00 AM	……
……	……
11:00 AM – 12:00 PM	Overlaps at noon 12:00 PM

Observing patterns from a data table works well here. There is another very interesting approach. Notice that whenever the minute hand meets the hour hand, except for the starting point, it means the minute hand has moved another one lap ahead of the hour hand. After the 12-hour period, the hour hand has finished only one complete lap, while the minute hand has completed twelve laps. To finish eleven laps more than the hour hand does, the minute hand must surpass (or meet) the hour hand eleven times. This is an exciting finding.

Angles on the Clock

[Example]

There are twelve numbers on an analog clock. How to calculate angle between the minute hand and the hour hand at the time when they both point straight to number(s)?

[Solution]
Because of 360 / 12 = 30, there is a 30-degree angle between every two numbers (representing hours) on the clock face.
When the time is at 6:00 (or 18:00), angle between minute hand and hour hand is:
    30 x 6 = 180 degrees
When the time is at 9:00 (or 21:00), smaller angle between minute hand and hour hand is:
    30 x (12 – 9) = 90 degrees
This applies to 3:00, or 15:00, too.

[Example]

The hour hand on a clock has moved 75 degrees since I last looked at the clock. How many hours have gone by? (2007 MIC)

[Solution]

12 (hours) * 75 (degrees) / 360 (degrees every 12 hours) = 2.5 hours

[Example]

The hands of an analog clock overlap each other at 12 o'clock. After the minute hand turns 960 degrees, what is the measure of the smaller angle between the hour hand and the minute hand? (2005 MathIsCool)

[Solution]

First step is to figure out how many minutes are equivalent to 960 degrees:

60 * (960 / 360) = 160 minutes, or 8/3 hours.

Second step is to compute how many degrees has the hour hand has advanced clockwise:

(360 / 12) * (8 / 3) = 80 degrees

Last step is to find out difference in degrees between the two hands:

(960 – 360 * 2) – 80 = 160 degrees

The smaller angle is then 180 – 160 = 20 degrees

Slow and Fast Clock

[Example]
This is a "slow clock" scenario:
A slow clock loses 15 minutes every hour. Suppose the clock is set to the correct time at 9 AM, what will be the correct time, when the slow clock first shows 10 AM?

[Solution]
When a normal clock moves an hour, the slow clock will move 45 minutes. The ratio between their constant speeds is 3:4. So when the slow clock moves an hour, the normal one will move 60 * (4 / 3) = 80 minutes after 9:00 a.m., which is 10:20 a.m.

[Example]

This is a "fast clock" scenario:

A fast clock gains 12 minutes every normal hour, and the fast clock shows the correct time at 1 PM. What is the correct time when the fast clock first registers 2 PM?

[Solution]

It tells us the moving speed between fast clock and normal one has a ratio as 72:60 = 6:5. If they both start at 1 p.m. at the same time, the normal clock will show 60 * (5 / 6) = 50 minutes after 1 p.m., which is 1:50 p.m., when the fast one is already 2 p.m.

Chasing on a Clock

[Example]

At what time the minute hand meets the hour hand during each hour period?

[Solution]

Let us look at it between one and two o'clock as an example. At just one o'clock, the hour hand is five units (five minutes) ahead of the minute hand. If one unit equals one minute's interval on a clock, the minute hand moves one unit per minute while the hour hand moves at the rate of 5/60 = 1/12 units per minute. It takes the minute hand 5 / (1 – 1/12) = 60/11 minutes to catch up with the hour hand. This is when both of two hands point at exactly the same position of one o'clock and 5.45 minutes, roughly speaking.

Here we have followed the formula as below, when both person A and person B go to the same direction and B is head of A:

Distance between A and B (B is ahead of A) = ([A's Speed] – [B's Speed]) * [Time-to-Meet]

Digital Clock

[Example]
How many times in a day is the "number" displayed on a digital 12-hour clock equal to a multiple of 12? E.g. at 1:20, the "number" is 120, and that is a multiple of 12.
(2016 MathIsCool)

[Solution]
Between 0 to 1 o'clock, 0:00, 0:12, 0:24, 0:36, 0:48 → 5 times;
Between 1 to 2 o'clock, 1:08, 1:20, 1:32, 1:44, 1:56 → 5 times;
Between 2 to 3 o'clock, 2:04, 2:16, 2:28, 2:40, 2:52 → 5 times;
Between 3 to 4 o'clock, 3:00, 3:12, 3:24, 3:36, 3:48 → 5 times;
Between 4 to 5 o'clock, 4:08, 4:20, 4:32, 4:44, 4:56 → 5 times;
....
Between 11 to 12 o'clock, 11:04, 11:16, 11:28, 11:40, 11:52 → 5 times;
....
5 * 12 * 2 = 120 times in total, during a day period.

[Summary]

- The hour hand advances one minute, as the minute hand moves every 12 minutes clockwise. Thus, ratio of the moving speed between them is 12:1, or 1: 1/12.

- For every round trip of a clock, the hour hand travels for 12 hours, equivalent to 360 degrees. When the minute hand travels 60 minutes (a.k.a. one hour), or 360 degrees, the hour hand will only move clockwise five minutes, but the clock itself moves one hour forward.

- Relation between a slow clock and a normal one has "[time diff. on normal clock] − [time diff. on slow clock] = [time diff. on normal clock] * [time lost rate]". Once we know ratio of "[time difference on normal clock] : [time difference on slow clock]", we can find out time read on slow clock or normal clock, provided one of them is known. Same rule applies to the fast clock.

- To calculate the time when both the hour hand and minute hand coincide between n and n+1 clock (n = 1, 2, ..., 11), you will need to know the total number of minutes' space (or units) between the hour hand and minute hand when it is n o'clock. Then you use the chasing formula, given speed difference between minute hand and hour hand (1 - 1/12) for every minute passed, to calculate the total minutes, the minute hand has to travel in order to meet the hour hand. This is exactly the position when both hands meet between n and n+1 o'clock.

[Practice Problems]

1. Mr. Bush's eighth grade class went on a field trip. They arrived at the park at 8:10 a.m. and left at 3:52 p.m. How much time did they spend at the park?

2. Suppose the time is now 2 o'clock on a twelve-hour clock, which runs continuously. What time will it show 1,000 hours from now?

3. A chime clock strikes 1 chime at one o'clock, 2 chimes at two o'clock, 3 chimes at three o'clock, and so forth. What is the total number of chimes the clock will strike in a twelve-hour period?

4. A light flashes every 3 seconds, a second light flashes every 4 seconds, and a third light flashes every 6 seconds. If all three lights flash together at 12 o'clock, what is the very next time on the clock that they will again flash together?

5. What is the number of degrees in the smaller angle between the hour hand and the minute hand on a clock that reads seven o'clock? (1989 AMC8)

6. What is the measure of the acute angle formed by the hands of the clock at 4:20 PM? (2003 AMC8)

7. It is 6:20 right now. What is the measure, in degrees, of the smaller angle between the minute and hour hand? (2016 MathIsCool)

8. A fast clock is set correctly at 12:00 noon, but it gains 4 minutes an hour. What will be the correct time when the fast clock next shows 12:00 midnight? (2009 MathIsCool)

9. Bailey's clock runs slow. For every minute of time that passes, her clock hands only move 40 seconds. If she sets her clock correctly at 3 PM, what is the correct time when her clock next shows 7:24 PM? (2012 MathIsCool)

10. Alice and Bob play a game involving a circle whose circumference is divided by 12 equally-spaced points. The points are numbered clockwise, from 1 to 12. Both start on point 12.

Alice moves clockwise and Bob, counterclockwise. In a turn of the game, Alice moves 5 points clockwise and Bob moves 9 points counterclockwise. The game ends when they stop on the same point. How many turns will this take? (2005 AMC8)

11. A clock has two hands, a short hand (hour) and a long hand (minute). Between eight and nine O'clock, what is the exact time do the two hands meet (round to the 100th minute)?

12. My 12-hour digital clock shows hours and minutes, but not seconds. The hour may have either 1 or 2 digits, but the minute always has 2 digits (with leading zeros used as necessary); for example, at 5 minutes after 7 PM, the clock displays"7:05". For how many minutes during a 12-hour period will my clock display no digit that isn't either a "1" or a "2"? (2013 MathIsCool)

13. A palindrome is a whole number that reads the same forwards and backwards. If one neglects the colon, certain times displayed on a digital watch are palindromes. Three examples are 1:01 4:44, and 12:21. How many times during a 12-hour period will be palindromes? (1988 AMC8)

Money

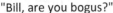

"Bill, are you bogus?"

[Basics] positive / negative, fraction, reciprocal number, rate of $/cent, US $ / other currencies

What are the last two digits of the result when 189,456,723 is multiplied by 17? Some people might attempt a complete multiplication between the two numbers, but may quickly realize that they do not really need to do so. A very similar scenario in our daily lives begins with you and your mom shopping at a grocery store. Your mom asks you to count the number of coins needed to buy 17 bottles of soymilk with a unit price at $2.23 including tax (yes, it's a little bit expensive). Are you going to calculate it by a "thorough" multiplication between 17 and 2.23? Not necessary. The quickest way is to ignore the $2 part, and focus on the result of 17 x $0.23, because only this part of the number matters for the coins. To answer the original question, what matters is the last two digits of the result are 17 and 23. "23" is the last two digits of number 189,456,723.

Common Sense and Basic Scenarios

You are suggested to work on the following as a warm-up exercise.

Some typical questions you may and you should ask are:
- How many different types of "money" do we know?
- We often use $1, $5, $10, and $20 bills. Do we have $2 and $50 bills?

Measurement and Conversion
- 1 **quarter** = 25 cents; 1 **dime** = 10 cents; 1 **nickel** = 5 cents; 1 **penny** = 1 cent
- Currency conversion
  - If US $1 equals to 120 Japanese Yen, 1 Yen = US$ ___; 500 Yen = US$ ___
  - If a product is for sale for RMB 10 dollars in China, then how much is it in US $? (US $1 = RMB 7 dollars)

Count coins

- A jar contains n nickels, d dimes, and q quarters. What is the ratio of nickels to total coins in the jar?
  (A) n / (d + q); (B) (d + q) / n; (C) n / (n * d * q); (D) d / (n + d + q); (E) n / (n + d + q)

Count bills
- John has a total of $300 in his wallet, in 10- and 20- dollar bills. If John has three times as many tens as twenties, how much money does he have in tens?
  (A) $40; (B) $60; (C) $120; (D) $180; (E) $360

Figure out price
- A large pizza costs $12 plus $0.75 for each topping. Which of the following equations represents the relationship between the price of a large pie, p, and the number of toppings it contains, t?
  (A) p = 12 + 0.75; (B) p = 12t + 0.75t; (C) p = 12 + 0.75t; (D) p = 12t + 0.75;
  (E) p = 0.75 * (12 + t)

Calculate money
- Patricia has $12 more than Rhoda and $15 more than Sarah. Together all three have $87. How much does Patricia have?

Transfer money
- Ann gave Betty as many cents as Betty had. Betty then gave Ann as many cents as Ann then had. At this point, each had 12 cents. How much did Ann have at the beginning?

Spend money
- Marissa bought 2 notebooks for $1.25 and 1 pencil for $.50. How much did she spend?

- Kevin wants to purchase a bicycle but is $23 short. Jerry wants to purchase the same bicycle but is $25 short. If they combine their money, they will have just enough to buy the bicycle. What is the cost of the bicycle?

Earn money
- Alice earned a total of $65 for working five days after school. Each day after the first day, she earned $2 more than she earned the day before. How much did she earn on the first day?

- One day, Carol bought apples at 3 for 25 cents and sold all of them at 2 for 25 cents. If she made a profit of $1 that day, how many apples did she sell?

Exchange between bill and coin
- Six dollars were exchanged for nickels and dimes. The number of nickels was the same as the number of dimes. How many nickels were there in the change?

- A dollar was changed into 16 coins consisting of just nickels and dimes. How many coins of each kind were in the change?

Exchange coins
- A person exchanged 390 pennies for quarters, dimes and nickels. The number of dimes in the exchange was twice the number of quarters and the number of nickels was twice the number of dimes. How many quarters were in the exchange?

Coin combination
- From a pile of 100 pennies (P), 100 nickels (N), and 100 dimes (D), select 21 coins which have a total value of exactly $1.00. In your selection you must also use at least one coin of each type. How many coins of each of the three types (P, N, D) should be selected?

- I have exactly ten coins whose total value is $1. If three of the coins are quarters, what are the remaining coins?

Buy stamps
- Carol spent exactly $1 for some 5 cents' stamps and some 13 cents' stamps. How many 5 cents' stamps did she buy?

[Example]
The average cost of a long-distance call in the USA in 1985 was 41 cents per minute, and the average cost of a long-distance call in the USA in 2005 was 7 cents per minute. Find the approximate percent decrease in the cost per minute of a long- distance call. (2007 AMC8)

[Solution]
The percent decrease is defined as (decreased cent amount / original cent amount).

$(41 - 7) / 41$ is roughly 80% decrease.

[Example]
The sales tax rate in Bergville is 6%. During a sale at the Bergville Coat Closet, the price of a coat is discounted 20% from its $90.00 price. Two clerks, Jack and Jill, calculate the bill independently. Jack rings up $90.00 and adds 6% sales tax, then subtracts 20% from this total. Jill rings up $90.00, subtracts 20% of the price, then adds 6% of the discounted price for sales tax. What is Jack's total minus Jill's total? (2005 AMC8)

[Solution]
Jack: $90 \times (1 + 6\%) \times (1 - 20\%)$
Jill: $90 \times (1 - 20\%) \times (1 + 6\%)$

As you see, they both have the same result. So, difference between their total is $0.

What we have learned from this example is, the order of taking discount off from the original price and adding tax to the price doesn't matter for the final price.

[Example]
Jamar bought some pencils costing more than a penny each at the school bookstore and paid $1.43. Sharona bought some of the same pencils and paid $1.87. How many more pencils did Sharona buy than Jamar? (2012 AMC8)

[Solution]
This is a common factor problem. 143 and 187 has one and only one common factor 11. So, Jamar bought $1.43 / $0.11 = 13 pencils, and Sharona bought $1.87 / $0.11 = 17 pencils, 4 more than Jamar.

[Practice Problems]

1. Kate had 5 coin purses, each with the same amount of money inside, but each with a different number of coins. If the total amount of money in all the purses was $1.40, what was the smallest possible total number of coins in all the purses? (2006 MathIsCool)

2. The membership fees for the gym consist of a monthly charge of $14 and a one-time new member fee of $16. Sam made a payment of $100 to pay his gym fees for a certain number of months, including the new member fee. How many months of membership did Sam include in his payment? (2015 MathIsCool)

3. Roger has 27 standard U.S. coins worth a total of $3.65. No coin is worth more than 30¢ or less than 3¢. How many dimes could Roger have? (2014 MathIsCool)

4. The Cool Math Club held a bake sale. They sold cakes for $3.50 each and pies for $6 each. They sold 24 items for a total of $119. The club donated 40 percent of the money earned from selling cakes to a charity. How many dollars did the club donate to charity? (2012 MathIsCool)

5. Holly earns $5 every day for taking care of her neighbor's cat. Every third day, she spends $6, but saves the rest of her earnings. After Holly was paid today, she spent $6, and then had $452 left in her savings. How many days from today will it take for her to have at least $1,200 in savings, assuming that she gets paid first on any day that she spends money? (2012 MathIsCool)

6. I have some dimes, which I can put into stacks with either 50 dimes in each stack, or 70 dimes in each stack, or 75 dimes in each stack, with no dimes left over in any case. In DOLLARS, what is the smallest possible value of my dime collection? (2011 MathIsCool)

7. Stacey buys a CD costing $19.48, and pays with a twenty-dollar bill. The cashier gives her change in dimes, nickels and/or pennies. How many different ways can the change be made? (2011 MathIsCool)

8. At the 2013 Winnebago County Fair a vendor is offering a "fair special" on sandals. If you buy one pair of sandals at the regular price of $50, you get a second pair at a 40% discount, and a third pair at half the regular price. Javier took advantage of the "fair special" to buy three pairs of sandals. What percentage of the $150 regular price did he save? (2013 AMC8)

9. The Amaco Middle School bookstore sells pencils costing a whole number of cents. Some seventh graders each bought a pencil, paying a total of 1.43 dollars. Some of the 30 sixth graders each bought a pencil, and they paid a total of 1.95 dollars. How many more sixth graders than seventh graders bought a pencil? (2009 AMC8)

10. In 2005 Tycoon Tammy invested 100 dollars for two years. During the first year her investment suffered a 15% loss, but during the second year the remaining investment showed a 20% gain. Over the two-year period, what was the change in Tammy's investment? (2008 AMC8)

11. Chandler wants to buy a 500-dollar mountain bike. For his birthday, his grandparents send him 50 dollars, his aunt sends him 35 dollars and his cousin gives him 15 dollars. He earns 16 dollars per week for his paper route. He will use all of his birthday money and all of the money he earns from his paper route. In how many weeks will he be able to buy the mountain bike? (2007 AMC8)

12. Business is a little slow at Lou's Fine Shoes, so Lou decides to have a sale. On Friday, Lou increases all of Thursday's prices by 10 percent. Over the weekend, Lou advertises the sale: "Ten percent off the listed price. Sale starts Monday." How much does a pair of shoes cost on Monday that cost 40 dollars on Thursday? (2003 AMC8)

13. A merchant offers a large group of items at 30% off. Later, the merchant takes 20% off these sale prices and claims that the final price of these items is 50% off the original price. The total discount is? (2002 AMC8)

14. You have nine coins: a collection of pennies, nickels, dimes, and quarters having a total value of $1.02, with at least one coin of each type. How many dimes must you have? (2000 AMC8)

15. Three generous friends, each with some money, redistribute the money as follow: Amy gives enough money to Jan and Toy to double each amount has. Jan then gives enough to Amy and Toy to double their amounts. Finally, Toy gives enough to Amy and Jan to double their amounts. If Toy had 36 dollars at the beginning and 36 dollars at the end, what is the total amount that all three friends have? (1998 AMC8)

16. At the grocery store last week, small boxes of facial tissue were priced at 4 boxes for $5. This week they are on sale at 5 boxes for$4. The percent decrease in the price per box during the sale was closest to? (1997 AMC8)

17. Ana's monthly salary was $2,000 in May. In June she received a 20% raise. In July she received a 20% pay cut. After the two changes in June and July, Ana's monthly salary was? (1996 AMC8)

18. A shopper buys a 100 dollar coat on sale for 20% off. An additional 5 dollars are taken off the sale price by using a discount coupon. A sales tax of 8% is paid on the final selling price. The total amount the shopper pays for the coat is? (1994 AMC8)

19. Jack had a bag of 128 apples. He sold 25% of them to Jill. Next he sold 25% of those remaining to June. Of those apples still in his bag, he gave the shiniest one to his teacher. How many apples did Jack have then? (1989 AMC8)

20. Sale prices at the Ajax Outlet Store are 50% below original prices. On Saturdays an additional discount of 20% off the sale price is given. What is the Saturday price of a coat whose original price is $180? (1986 AMC8)

21. Using only pennies, nickels, dimes, and quarters, what is the smallest number of coins Freddie would need so he could pay any amount of money less than a dollar? (2010 AMC8)

Travel

[Basics] positive / negative number, fraction, ratio, average, time calculation

The story of "The Hare and the Tortoise" may be very familiar to you. We have learned a lot from the hare's failure. If the hare had not slacked off until after it reached the finish line, he would have undoubtedly won the race. However, the problem was not whether the hare slacked off or not. The hare's fault was that he slacked off for longer than he should have. An experienced racer knows at what timing, and at what point he should run hard. He also knows it is not a good idea to run at full speed at all times during the race. A good racer plans for when and where in the course to slow down, to take a little "rest" in order to save energy for later. As long as he will eventually win, it does not really matter if he wins by a large or a little margin. If the hare had calculated time better, he would not have lost the race to the tortoise.

[Question]

If the hare runs 10 units per second, which is 10 times faster than the tortoise does; and the whole race is a 9000-units long course, do you know how many minutes the hare can "slack" in the middle of the course and still win the race?

[Answer]

This is a typical travel problem, the three key elements are distance, time, and speed. They follow the formula of d = v * t, or, Distance = Velocity * Time.

If the hare and the tortoise keep running from the beginning to the end at their constant speeds, the hare can reach the goal by 9000 / 10 = 900 seconds, and the tortoise will need 9000 / 1 = 9000 seconds to finish the race. This means the hare can take a rest for up to 9000 – 900 = 8100 seconds, or135 minutes. The hare did have lots of time to slack.

The story had a new development afterwards. The hare and the tortoise had another race. Half of the race route was on a river. Since the hare had no idea how to swim across the river, the hare lost the game again to the tortoise. The problem for the hare this time was not the river itself. The hare could have found whatever way to across it, for example boating or taking a ferry, even if it moves very slowly on the river.

[Question]

The whole race has a 9000 units' long distance, half of which is on a river. On the land, the hare runs 10 units per second, which is 10 times faster than the tortoise's speed. On the river, the tortoise can swim 10 units per second, while the hare can only manage to move one unit per second by a boat. Can you tell who will win the race this time, assuming they both are serious to compete without slacking during the race?

[Answer]

To complete the whole race,

- the hare takes time: 4500/10 on the land and 4500/1 on the river

    => 4950 seconds in total

- the tortoise takes time: 4500/1 on the land and 4500/10 on the river

    => 4950 seconds in total

It would take exactly the same time for both of them to finish the race.

In a general travel problem, the key thing is to identify distance, time, velocity involved in the scenario. Sometimes, it requires computation (i.e. addition, subtraction, multiplication, division) to combine partial distance, or time spent, or total speeds. Sometimes, it assumes one of the three elements is fixed, and it requires us to work around the relation between the other two elements.

For example, if time is a constant, then the distance travelled is proportional to one's speed (a.k.a. velocity). The same rule applies to when one's speed is a constant; if total distance is a fixed number, then one's speed and the time spent on the travel are inversely proportional.

One-way Trip
       Three Elements: $d = v * t$
       If A runs 1 mile for 1 hour, then
       (a) how many miles A can run for 2 hours?
       (b) how long A can run for 3 miles?

A freight train travels 1 mile in 1 minute 30 seconds. How many miles will the train travel in 1 hour at this rate?

Round-Trip

Speed = 2 * Distance / Total Time

(Considering the wind's speed, it would become a more completed scenario.)
A motorist made a 60-mile trip averaging 20 miles per hour. On the return trip, he averaged 30 miles per hour. What was the motorist's average speed for the entire trip?

A Chases B
Race scenario - Associate with clock / two hands meet problem

A Meets B
The train Henry leaves city A at the exactly the same time when another train Gordon on the same route leaves city B. Train Henry is moving twice as fast as train Gordon. The route between city A and city B is 240 miles long. How far is the point from city A when both trains meet?

Track Racing
- Two dogs run around a circular track 300 feet long. One dog runs at a steady rate of 15 feet per second. The other dog's steady speed is 10 feet per second. Suppose they start at the same point and same time. What is the least number of seconds that will elapse before they are again together at the starting point?

Other Scenarios
- Spare tire's trip
  The 5 tires of a car (4 + 1 spare) were each used equally on a car that had traveled 60,000 miles. How many miles each tire was used for during that time?

- Scheduling
  Peter had a 12:00 noon appointment that was 60 miles from his home. He drove from his home at an average rate of 40 miles per hour and arrived 15 minutes late. At what time did Peter leave home for the appointment?

- Up and down in elevator

    Suppose you enter an elevator at a certain floor. Then the elevator moves up 6 floors, down 4 floors, and up 3 floors. You are then at floor 7. At which floor did you initially enter the elevator?

[Example]
Two dogs 50 miles apart are racing towards each other. One is running at 10 miles per hour and the other at 15 miles per hour. While the dogs are running towards each other, a flea jumps back and forth between their noses at 40 miles per hour. How many miles does the flea jump before being squished between the dogs' noses? (2005 MathIsCool)

[Solution]
Flea's jumping speed is already given as a constant number 40 mph. If we know the time, we will be able to use the $d = v * t$. Notice the time the flea spent on this trip is the same time as the two dogs running towards each other until they meet at somewhere between them. Now the question is how long does it take the two dogs to meet?
It can be solved by 50 (miles) / (10 + 25) (miles per hour) = 2 hours.
Therefore, the flea will jump 10 (mph) * 2 (hours) = 20 miles.

Travel problem would be a good challenge to undermine relations between distance and other dependency factor, such as mileage and gasoline.

[Example]
Karl's car uses a gallon of gas every 35 miles, and his gas tank holds 14 gallons when it is full. One day, Karl started with a full tank of gas, drove 350 miles, bought 8 gallons of gas, and continued driving to his destination. When he arrived, his gas tank was half full. How many miles did Karl drive that day? (2016 AMC8)

[Solution]
After Karl drove 350 miles, his car has used 350 / 35 = 10 gallons. After have added 8 gallons, the car had 14 - 10 + 8 = 12 gallons left in the tank. When Karl arrived at his destination, the car used up 12 - 14/2 = 5 gallons. Therefore, Karl drove 350 + 35 * 5 = 525 miles in total on that day.

[Summary]
- Three key elements – Distance, Time, Speed

- One-way trip scenario: Speed = Distance / Time
    - or   Distance = Speed * Time
    - or   Time = Distance / Speed

- Round-trip scenario: Speed = 2 * Distance / Time (total)

- A meets B scenarios:
  (a) A and B go toward each other
       Distance (between A and B) = (A-Speed + B-Speed) * Time-to-Meet
  (b) A and B go to the same direction, and B is ahead of A
       Distance (that B is ahead of A) = (A-Speed − B-Speed) * Time-to-Meet

[Practice Problems]

1. On a dark and stormy night, Snoopy suddenly saw a flash of lightning. Ten seconds later he heard the sound of thunder. The speed of sound is 1088 feet per second and one mile is 5280 feet. Estimate, to the nearest half-mile, how far Snoopy was from the flash of lightning. (2001 AMC8)

2. The results of a cross-country team's training run are graphed below. Which student has the greatest average speed? (2005 AMC8)

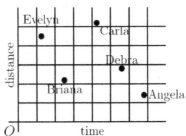

3. At the same time, Stuart and Theo start walking toward each other along a number line. Stuart starts at point 78 and walks at a speed of 4 units per minute. Theo starts at point 14 and walks at 3 units per minute. When Theo reaches point 95, how many units past 0 will Stuart have walked? (2013 MathIsCool)

4. Biff is riding his bike from his house to Eho's house to study math. Biff has ridden 8 miles plus two-thirds of the total distance, and still has 5 miles to go. How many miles is it from Biff's house to Eho's house? (2009 MathIsCool)

5. Randy and Alex are racing bikes around a circular track 800 meters long. If Randy bikes one and one-fourth times as fast as Alex, but Alex has a 500 meter head start, how many laps will it take for Randy to catch up with Alex? Assume that Randy bikes at a rate of 8 meters per second, and that Randy and Alex start at the same time and bike in the same direction. If your answer is not a whole number, give it as a decimal. (2009 MathIsCool)

6. A snail tries to get out of a well. Each day it climbs up the side of the well 4 feet and each night it slides down the well 2 feet and 6 inches. If the snail starts 40 feet down inside in the morning, how many days will the snail take to get out of the well? (2008 MathIsCool)

7. Annie and Bonnie are running laps around a 400-meter oval track. They started together, but Annie has pulled ahead because she runs 25% faster than Bonnie. How many laps will Annie have run when she first passes Bonnie? (2016 AMC8)

8. A straight one-mile stretch of highway, 40 feet wide, is closed. Robert rides his bike on a path composed of semicircles as shown. If he rides at 5 miles per hour, how many hours will it take to cover the one-mile stretch? Note: 1 mile= 5280 feet (2014 AMC8)

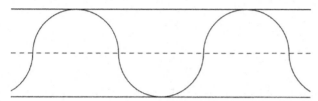

9. The Incredible Hulk can double the distance he jumps with each succeeding jump. If his first jump is 1 meter, the second jump is 2 meters, the third jump is 4 meters, and so on, then on which jump will he first be able to jump more than 1 kilometer? (2013 AMC8)

10. Ted's grandfather used his treadmill on 3 days this week. He went 2 miles each day. On Monday he jogged at a speed of 5 miles per hour. He walked at the rate of 3 miles per hour on Wednesday and at 4 miles per hour on Friday. If Grandfather had always walked at 4 miles per hour, he would have spent less time on the treadmill. How many minutes less? (2013 AMC8)

11. As Emily is riding her bicycle on a long straight road, she spots Emerson skating in the same direction ½ mile in front of her. After she passes him, she can see him in her rear mirror until he is ½ mile behind her. Emily rides at a constant rate of 12 miles per hour, and Emerson skates at a constant rate of 8 miles per hour. For how many minutes can Emily see Emerson? (2010 AMC8)

12. Austin and Temple are 50 miles apart along Interstate 35. Bonnie drove from Austin to her daughter's house in Temple, averaging 60 miles per hour. Leaving the car with her daughter, Bonnie rode a bus back to Austin along the same route and averaged 40 miles per hour on the return trip. What was the average speed for the round trip, in miles per hour? (2009 AMC8)

13. Cassie leaves Escanaba at 8:30 AM heading for Marquette on her bike. She bikes at a uniform rate of 12 miles per hour. Brian leaves Marquette at 9:00 AM heading for Escanaba on his bike. He bikes at a uniform rate of 16 miles per hour. They both bike on the same 62-mile route between Escanaba and Marquette. At what time in the morning do they meet? (2006 AMC8)

14. Gage skated 1 hr 15 min each day for 5 days and 1 hr 30 min each day for 3 days. How long would he have to skate the ninth day in order to average 85 minutes of skating each day for the entire time? (2002 AMC8)

15. The third exit on a highway is located at milepost 40 and the tenth exit is at milepost 160. There is a service center on the highway located three-fourths of the way from the third exit to the tenth exit. At what milepost would you expect to find this service center? (1999 AMC8)

16. Buses from Dallas to Houston leave every hour on the hour. Buses from Houston to Dallas leave every hour on the half hour. The trip from one city to the other takes 5 hours. Assuming the buses travel on the same highway, how many Dallas-bound buses does a Houston-bound bus pass in the highway (not in the station)? (1995 AMC8)

17. On a trip, a car traveled 80 miles in an hour and a half, then was stopped in traffic for 30 minutes, then traveled 100 miles during the next 2 hours. What was the car's average speed in miles per hour for the 4-hour trip? (1992 AMC8)

18. At the beginning of a trip, the mileage odometer read 56,200 miles. The driver filled the gas tank with 6 gallons of gasoline. During the trip, the driver filled his tank again with 12 gallons of gasoline when the odometer read 56,560. At the end of the trip, the driver filled his tank again with 20 gallons of gasoline. The odometer read 57,060. To the nearest tenth, what was the car's average miles-per-gallon for the entire trip? (1986 AMC8)

19. A ball is dropped from a height of 3 meters. On its first bounce it rises to a height of 2 meters. It keeps falling and bouncing to 2/3 of the height it reached in the previous bounce. On which bounce will it rise to a height less than 0.5 meters? (2008 AMC8)

20. A ball with diameter 4 inches starts at point A to roll along the track shown. The track is comprised of 3 semicircular arcs whose radii are $R_1 = 100$ inches, $R_2 = 60$ inches, and $R_3 = 80$ inches, respectively. The ball always remains in contact with the track and does not slip. What is the distance the center of the ball travels over the course from A to B? (2013 AMC8)

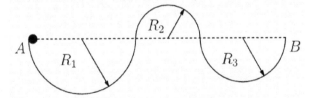

21 Each day for four days, Linda traveled for one hour at a speed that resulted in her traveling one mile in an integer number of minutes. Each day after the first, her speed decreased so that the number of minutes to travel one mile increased by 5 minutes over the preceding day. Each of the four days, her distance traveled was also an integer number of miles. What was the total number of miles for the four trips? (2017 AMC 8)

Four-legs vs. Two-legs

[Basics] ratio and number operations

Have you been challenged with this kind of problem before? "There are 20 chickens and rabbits in total inside a cage. If the total number of legs they have is 56, how do you know the number of chickens?"

Imagine this, if all of them in the cage were chickens, how many legs in total would they have then? It would be 20 x 2 = 40 legs, since each chicken has two legs. But the actual total number of legs is 56, where do the extra 16 legs come from?

Continue to imagine, if we "replace" one chicken with one rabbit, we will "gain" two more legs. To make up for the 16 legs' difference, we would need to "replace" 16/2 = 8 chickens with rabbits. There are 8 rabbits and 20 − 8 = 12 chickens are in the cage.

Does the idea of imagination and swapping method seem new to you? Let's look at another example to get familiar with the way of thinking.

[Question]
If there are 19 coins including quarters and pennies, and the total value of them is just 91 cents, how many quarters do you think it includes?

[Answer]
If all of the coins were pennies, then the difference in the sum would be 91 − 19 x 1 = 72 cents. To make up for the difference, we need to "replace" 72 / (25-1) = 3 pennies with quarters, so the answer is three quarters.

One more common scenario in the same category as counting coins is about buying stamps. Since there are many types of stamps with different face values, the idea introduced above is 100% applicable when you count the stamps.

[Practice Problems]

1. Lou bought five pieces of fruit. Each piece of fruit was either an apple, weighing ½ pound, or an orange, weighing ¾ pound. The total weight of the fruit was a whole number of pounds. If apples cost 80¢ a pound and oranges cost $1.10 a pound, how much did Lou's fruit cost, as a decimal number of dollars? (2010 MathIsCool)

2. Tickets on a bus were $4.00 and $6.00. A total of 45 tickets were sold and $230 earned. How many $4.00 tickets were sold? (2007 MathIsCool)

3. Anne's Game Farm has only ten-point bucks and six-point bucks. If there are a total of 87 bucks with a total of 602 points between them, how many six-point bucks are there? (2007 MathIsCool)

4. Cherie and Katie were looking at elephants and ostriches in the Children's Zoo. Cherie saw 12 heads, and Katie saw 30 legs. How many elephants did they see? (2007 MathIsCool)

5. There are 19 heads and 62 legs in a field of ducks and camels. How many camels are in the field? (2006 MathIsCool)

6. Richa and Yashvi are going to Jamaica with their school. They plan on attending a fair where the admission for children is $1.50 and $4.00 for adults. On a specific day, 2200 people enter the fair and $5,050 is collected. How many children attended? (2017 MathIsCool)

7. A chair has 4 legs, a stool has 3 legs and a table has 1 leg. At a birthday party, there are 4 chairs per table and a total of 18 pieces of furniture. One of the children counts 60 legs total. How many stools are there? (2016 MathIsCool)

8. The Fort Worth Zoo has a number of two-legged birds and a number of four-legged mammals. On one visit to the zoo, Margie counted 200 heads and 522 legs. How many of the animals that Margie counted were two-legged birds? (2012 AMC8)

9. A group of children riding on bicycles and tricycles rode past Billy Bob's house. Billy Bob counted 7 children and 19 wheels. How many tricycles were there? (2003 AMC8)

10. In a mathematics contest with ten problems, a student gains 5 points for a correct answer and loses 2 points for an incorrect answer. If Olivia answered every problem and her score was 29, how many correct answers did she have? (2002 AMC8)

11. On a twenty-question test, each correct answer is worth 5 points, each unanswered question is worth 1 point and each incorrect answer is worth 0 points. Which of the following scores is **NOT** possible? (2001 AMC8)
    (A) 90   (B) 91   (C) 92   (D) 95   (E) 97

12. A multiple choice examination consists of 20 questions. The scoring is +5 for each correct answer, -2 for each incorrect answer, and 0 for each unanswered question. John's score on the examination is 48. What is the maximum number of questions he could have answered correctly? (1987 AMC8)

Age

[Basics] ratio

Who ages faster, my dad or I? This is an interesting question. If my dad's age is thirty, which is three times older than I am, how many years later will my age become half of my dad's?

The question's sentiment implies that my dad ages slower than I do. Is it possible? Let us take a close look at what is going on here. My dad is now 30 and I am 10, so next year my dad becomes 31 and I will be 11. The next of the next year my dad will be 32 and I will be 12, and so on. In five years, my dad's age will become 35 and I will be 15 years-old. What do you see if you compare the ratio of my dad's age against mine? It is: (30/10) > (31/11) > (32/12) > (33/13) > (34/14) > (35/15). Yes, the ratio decreases, although my dad's age and my age both increase by one each year. However, to keep the same ratio (30/10) as I grow, my dad would need to "grow" three at next year and so forth, because 30/10 = (30 + 3) / (10 + 1) = (30 + 6) / (10 + 2) = ...

Can I say that my dad grows in age slower than I do? Well, yes and no. When we look at the ages, my dad and I grow at the same "speed" each year - one per year. When we look at the ratio as mentioned above, though, it looks like that I grow faster than my dad. When we consider the growth rate, what we get is that next year my dad's "growth" will be 31/30 and mine will be 11/10. This clearly indicates my "growth" rate (11/10) is higher than my dad's (31/30). It is simply because the ratio, or the rate of growth, is calculated based on something, or a base number. It is a relative number. It may not be an appropriate indicator to describe how fast my dad and I grow. However, it is helpful to analyze the trend of growth and associate them with related math concepts in order to understand them in-depth.

29

To solve the original problem, you can always adopt the methodology of algebra. But, if you are not ready for it, an alternative approach is to draw two horizontal line segments, which represent your dad's age and your age respectively on a piece of paper. Then, you figure out how to extend the two lines by the same size.

On the paper, you analyze the number relation between the two different line segments. Notice the "?" represents the same length, or same number of years. It becomes clear that (30 − 10 x 2 = 10) is the right formula to reach the correct answer. In ten years, my dad will be forty and I will be twenty.

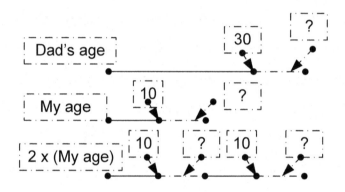

[Example]
The sum of the ages of Al and Bill is 25; the sum of the ages of Al and Carl is 20; the sum of the ages of Bill and Carl is 31. Who is the oldest of the three, and how old is he?

[Solution]
When you draw something on your scratch paper as below, you may be able to realize (25 + 20 + 31) = 76 is actually twice the total ages of Al, Bill, and Carl. Their total age is then 76 /2 = 38. Now, it is easy to find that Al's age = 38 − 31 = 7; Bill's = 38 − 20 = 18; Carl's = 38 − 25 = 13. Therefore, the oldest of the three is Bill, who is 18 years old.

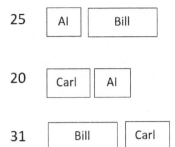

[Example]

Fric is four times as old as Frac. Four years ago, Fric was ten times as old as Frac. What is Fric's current age?

[Solution]

As depicted in graph, Fric's current age minus four is 10 times Frac's current age minus four, while Frac's current age is one-fourth of Fric's current age.

Dividing one line representing Frac's current age into three equal parts (see vertical line separators), we find that 4 is actually twice as long as Frac's age four year before. This means Frac was 2 years old four years ago. Frac is now 6 years old, which concludes that Frac's current age is 24.

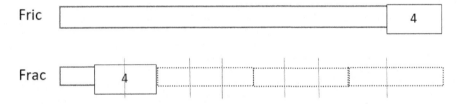

[Summary]

Scenario Patterns

1. If you know the sum and difference between A's age and B's age, how do you figure out A's and B's age?

2. If you know the sum of A's age and B's age, and you also know the ratio between their ages, how do you figure out A's and B's age?

3. If you know A's age and B's age, how many years later A's age will become N times as old as B's age? N is a positive integer.

Key: Draw lines and analyze their relations in length.

[Practice Problems]

1. Ten years ago, Hanson's age was half of the age he will be in 16 years. How old is he now?

2. In 20 years, Henry's age will be five times as old as he is today. What is his current age?

3. The age of a man is the same as his wife's age with the digits reversed. The sum of their ages is 99 and the man is 9 years older than his wife. How old is the man?

Work

[Basics] ratio, travel problem,

When one person does a job at a constant rate, a.k.a. work efficiency, you can calculate the time that person is able to accomplish the job. This is an equivalent approach to solve simple travel problems. The difference is, the rate is called "speed" in travel scenarios, while it means "efficiency" in work scenario.

[Example]
Amy can mow 600 square yards of grass in 1.5 hours. At this rate, how many minutes would it take her to mow 900 square yards?

[Solution]
90 x 900 / 600 = 135 minutes

When several people work together on a project, each person's work efficiency contributes to work progress, as well as final outcome. When the total work volume is unclear to us, we may make an assumption of the whole workload as "1", to simplify computation. You will find the result is the same if you assume the workload equal to any other number.

The bath tub scenario is another one that also belongs to work problem.

[Example]
The cold water faucet of a bath tub can fill the tub in 15 minutes. The drain, when opened, can empty the full tub in 20 minutes. Suppose the tub is empty, and the faucet and drain are both opened at the same time, how long will it take to fill the tub?

[Solution]
1 / (1/15 − 1/20) = 60 minutes

[Summary]
Try working on these examples under each pattern.

Pattern-1
One job needs X number of people, Y days (or hours) to finish. How many days does the same job require, if there is N number of people available? (assuming that everyone has the same working efficiency, a.k.a. work efficiency rate: the same work result per day)
(Individual rate) * X * Y = (Individual rate) * N * <The number of days the question is asking for>
    Obviously, X * Y / N is the answer key.

[Example]
A work crew of 3 people requires 3 weeks and 2 days to do a certain job. How long would it take a work crew of 4 people to do the same job if each person of both crews works at the same rate as each of the others?

Pattern-2
One can do the job for X hours, while the other can do the same job for Y hours. What if both of them work together?

- Define the whole job as "1" (the simpler the better!), their working rates are 1/X and 1/Y separately. Then, the problem becomes Motion / "A meets B" scenario.

[Example]
Alice needs 1 hour to do a certain job. Betty, her elder sister, can do the same job in ½ hour. How many minutes will it take them to do the job if they work together at the given rates?

Pattern-3 (Extension of Pattern-2)
If you know one person's work efficiency rate and the combined rate for two, How to figure out the other person's rate? (Extension of Pattern-2)

[Example]
Greg and Larry are working together to sand and refinish the hard-wood floor in a room. It takes them 18 hours to complete the job working together. If it takes Greg 30 hours to do the job alone, how many hours does it take Larry to do the job alone?

[Practice Problems]

1. A work team of four people completes half of a job in 15 days. How many days will it take a team of ten people to complete the remaining half of the job? Assuming that each person of both teams works at the same rate.

2. Homer began peeling a pile of 44 potatoes at the rate of 3 potatoes per minute. Four minutes later Christen joined him and peeled at the rate of 5 potatoes per minute. When they finished, how many potatoes had Christen peeled? (2001 AMC8)

3. Adam can run a lap on a certain circular track in 50 seconds. Grampy Sampy can run a lap on this track in 90 seconds. They start at the same location at the same time and move in the same direction. If they each run at a constant speed, how many seconds will it take before Adam is next even with Grampy Sampy? If your answer is not a whole number, give it as a decimal. (2012 MathIsCool)

4. George walks 1 mile to school. He leaves home at the same time each day, walks at a steady speed of 3 miles per hour, and arrives just as school begins. Today he was distracted by the pleasant weather and walked the first ½ mile at a speed of only 2 miles per hour. At how many miles per hour must George run the last ½ mile in order to arrive just as school begins today? (2014 AMC8)

5. Niki usually leaves her cell phone on. If her cell phone is on but she is not actually using it, the battery will last for 24 hours. If she is using it constantly, the battery will last for only 3 hours. Since the last recharge, her phone has been on 9 hours, and during that time she has used it for 60 minutes. If she doesn't talk any more but leaves the phone on, how many more hours will the battery last? (2004 AMC8)

6. A birdbath is designed to overflow so that it will be self-cleaning. Water flows in at the rate of 20 milliliters per minute and drains at the rate of 18 milliliters per minute. One of these graphs shows the volume of water in the birdbath during the filling time and continuing into the overflow time. Which one is it? (2002 AMC8)

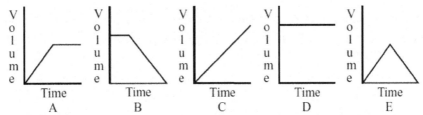

7. Joe had walked half way from home to school when he realized he was late. He ran the rest of the way to school. He ran 3 times as fast as he walked. Joe took 6 minutes to walk half way to school. How many minutes did it take Joe to get from home to school? (2005 AMC8)

Mixture

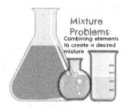

[Basics] percentage, ratio, number computation

When you find that a soup is too salty, what do you do? Removing a little bit of salt from the soup does not work realistically in this case. You can add some water to dilute it, though. Of course, we do not need to compute the volume of water we add at that time. We may just adjust the volume and taste it until we feel comfortable with it. However, when we work on a science project, we definitely need to figure out the accurate measurement.

[Example]
A mixture of 30 liters of paint is 25% red tint, 30% yellow tint, and 45% water. Five liters of yellow tint are added to the original mixture. What is the percent of yellow tint in the new mixture? (2007 AMC 8G/17)

[Solution]
Let's build a table with the information filled in to understand what data is available, and what is we need to find out.

	Red tint	Yellow tint	Water	Total
Before	25%, 7.5 L	30%, 9 L	45%, 13.5 L	30 L
After		? %, 14 L		35 L

30% of the original 30 liters of paint was yellow ➜ 9 liters of paint were yellow
After 5 liters of yellow paint were added to make the new mixture, ➜ There are 9 + 5 = 14 liters of yellow tint in the new mixture.
5 liters of paint were added to the original 30 ➜ There are now a total of 35 liters of paint in the new mixture.
14 out of 35 is 40% ➜ There is 40% of yellow tint in the new mixture.

If you remember, the following problem is one of the practice problems in the earlier chapter "Four-legs vs two-legs". This problem may be solved by type of "mixture problem" scenario as well.

[Example]

Tickets on a bus were $4.00 and $6.00. A total of 45 tickets were sold and $230 earned. How many $4.00 tickets were sold? (2007 MathIsCool)

[Solution]

	Tickets sold	$ per ticket	Total
$4 tickets	?	$4	? x $4
$6 tickets	45 - ?	$6	(45 - ?) x $6
Total	45		$230

? x 4 + (45 - ?) x 6 = 45 x 6 - ? x 2 = 230, which is, 270 – 230 = ? x 2 ➜ ? = 20 $4 tickets.

[Practice Problems]

1. Two 600 mL pitchers contain orange juice. One pitcher is 1/3 full and the other pitcher is 2/5 full. Water is added to fill each pitcher completely, then both pitchers are poured into one large container. What fraction of the mixture in the large container is orange juice? (2004 AMC8)

2. Miki has a dozen oranges of the same size and a dozen pears of the same size. Miki uses her juicer to extract 8 ounces of pear juice from 3 pears and 8 ounces of orange juice from 2 oranges. She makes a pear-orange juice blend from an equal number of pears and oranges. What percent of the blend is pear juice? (2002 AMC8)

3. Ten grams of sugar are added to a 40-g serving of a breakfast cereal that is 30% sugar. What is the percent concentration of sugar in the resulting mixture?

Painted Cubes

[Basics] cube, rectangle

A human being's brain is theoretically separated into two parts, the left brain and right brain, utilized for different thinking. Generally, the right brain focuses on feeling, creativity, etc., while the left brain is used for thinking that is logical, objective, and analytical. Which part of the brain mainly supports our capabilities of spatial problem solving? Usually it is said to be the left brain. However, different people might have different role "pre-designed" already in their left brain and right brain, in the same way as there are "right-handed" and "left-handed" people around. Moreover, the spatial thinking process involves not only imagination but also logical analysis and synthesis as well. Although there is not enough academic evidence to prove this, it is more likely that one's spatial awareness is a skill accomplished by the "whole-brained" effort. The following problem is a typical spatial thinking test.

[Question]

A huge cube with 10 units on each side is painted. It is then cut into small cubes with 1 unit on each of its sides. How many small cubes are painted on at least one side?

[Solution]

A very basic approach is to figure out the unknown data from the available information. Since the total number of the small cubes is 10 x 10 x 10 = 1000, the problem can be narrowed down to figure out how many small cubes are not painted at all?

Through a spatial thinking process, we try to find out the invisible object(s) based on the visible one(s). All of the small cubes that are painted on at least one side should be visible; otherwise, there is no way to touch and paint them. In other words, all of the small cubes that cannot be painted are the invisible cubes. On the other hand, it is possible to count all of the visible small cubes, which are on the six surfaces of the huge cube. However, this is not the easiest way.

Looking into the huge cube, we discover that all of the invisible small cubes are actually hidden inside a big cube with $10 - 2 = 8$ units on each side, after removing all of the small cubes located on the surfaces of the six sides, including those cubes at the edges and the corners. Therefore, the correct answer to the question will be $1000 - 8 \times 8 \times 8 = 488$.

[Example]
12 cubes are placed together to form a rectangular solid two cubes by two cubes by three cubes. If the entire outside of the rectangular solid is painted and the solid is taken apart into the 12 individual cubes, what is the total number of faces of the cubes that remain unpainted? (2004 MathIsCool)

[Solution]
Pay attention to the word "faces", not "cubes".
Total number of faces is $12 \times 6 = 72$.
Total number of faces painted is $2 \times (2 \times 3 + 2 \times 3 + 2 \times 2) = 32$.
Then, the total number of faces unpainted is $72 - 32 = 40$.

[Example]
If a 6 by 6 by 6 cube is made up of 1 by 1 by 1 cubes and the outside is painted orange, how many cubes will have only 2 orange sides? (2003 MathIsCool)

[Solution]
Think about where these cubes exist,
$(6 - 2) \times 4 \times 3 = 48$ cubes

[Example]
Tony has a cube eight inches on a side. He paints the entire outside of the cube. Next, he cuts the cube into smaller cubes 1 inch on a side, how many smaller cubes will have paint on only one side? (2001 MathIsCool)

[Solution]
$(8 - 2) \times (8 - 2) \times 6 = 216$ cubes

[Example]
A cube with 3-inch edges is to be constructed from 27 smaller cubes with 1-inch edges. Twenty-one of the cubes are colored red and 6 are colored white. If the 3-inch cube is constructed to have the smallest possible white surface area showing, what fraction of the surface area is white? (2014 AMC8)

[Solution]
For the least possible surface area, we should have 1 cube in the center, and the other 5 with only 1 face exposed. This gives 5 square inches of white surface area. Since the cube has a surface area of 54 square inches, our answer is 5/54.

[Practice Problems]

1. A cube with 3-inch edges is made using 27 cubes with 1-inch edges. Nineteen of the smaller cubes are white and eight are black. If the eight black cubes are placed at the corners of the larger cube, what fraction of the surface area of the larger cube is white? (2006 AMC8)

2. A cube has edge length 2. Suppose that we glue a cube of edge length 1 on top of the big cube so that one of its faces rests entirely on the top face of the larger cube. What is the percent increase in the surface area (sides, top, and bottom) from the original cube to the new solid formed is closest to? (2000 AMC8)

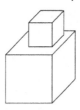

3. A cube three inches on an edge is made from white unit cubes, each one inch on an edge. I want to paint as many of the unit faces blue as possible, but no two blue unit faces can share a side. It's OK for blue unit faces to touch diagonally, however. (See the diagrams for examples.) Find the largest number of unit faces of this cube that I can paint blue. (2009 MathIsCool)

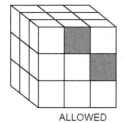
ALLOWED

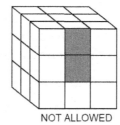
NOT ALLOWED

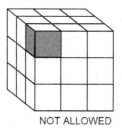
NOT ALLOWED

4. Marla has a large white cube that has an edge of 10 feet. She also has enough green paint to cover 300 square feet. Marla uses all the paint to create a white square centered on each

face, surrounded by a green border. What is the area of one of the white squares, in square feet? (2012 AMC8)

5. A 4 x 4 x 4 cubical box contains 64 identical small cubes that exactly fill the box. How many of these small cubes touch a side or the bottom of the box? (1998 AMC8)

6. A 2 by 2 square is divided into four 1 by 1 squares. Each of the small squares is to be painted either green or red. In how many different ways can the painting be accomplished so that no green square shares its top or right side with any red square? There may be as few as zero or as many as four small green squares. (1994 AMC8)

7. An artist has 14 cubes, each with an edge of 1 meter. She stands them on the ground to form a sculpture as shown. She then paints the exposed surface of the sculpture. How many square meters does she paint? (1989 AMC8)

Spatial Thinking

Pots in the kitchen

Pots are used to cook food and contain food. There are various kinds of pots in a kitchen. A pan is certainly one of the many pots in the kitchen. As a food container, a bowl may be thought as a special type of pot. Have you discovered that almost all of the pots are basically in a round shape on the surface? We generally prefer to design a product with a symmetric shape. Other than circles, there are many different kinds of symmetric shapes, such as equilateral triangles, squares, rhombuses, etc. Are there special reasons that we have pots in a round shape design? There may be several good answers to this question.

Reason 1

Pots have lids to cover them. The pot cover preserves the temperature when the pot is in cooking condition. It should completely cover the surface of the pot. Considering the material and the outlook, the pot cover is better not to be made too big in size. Another very important thing is, the pot cover should not drop down into the pot! A round shape is the best fit for this purpose. On the contrary, if we use a pot cover in a square shape, it can easily find its position to drop into the pot. The same results occur when it is designed by other shapes.
(Note: A very well-known job interview question – "why is a man hole usually in a round shape?" has indeed the same answer as above.)

Reason 2

The round shape is "center-oriented". When you have a pot in a round shape design, it is easy to warm the food inside the port and it is convenient when you try to clean out the stuff from inside the pot. However, if the pot were in other shapes such as a rectangle prism which has the surface in a rectangle or a square shape, it would bring inconvenience when you are trying to make sure all the food inside the pot is equally heated, let alone how much trouble it would cause when you want to clean the pot from inside.

Reason 3

A pot in a round shape is also round at its outside, which is much safer than any other shape. If the pot is in a square shape, since it has sharp edges towards outside, it will be dangerous when it happens to hit a kid's body or a fragile object.

Every single object has its own shape. The design for the shape of every object has its own reason. Understanding why they have to be in a specific shape is extremely helpful to realize the unique attributes of the geometrical shapes and their applicable circumstances around us.

[Practice Problems]

1. A solid hemisphere of radius 10 cm is glued to a table with its flat side down. Two ants, Ann and Andy, travel from point A to point B. Ann stays on the table and Andy goes over the top of the hemisphere, each taking the shortest path for her or her route. How many cm more does Andy travel than Ann? (2008 MathIsCool)

2. A booklet is made by folding and stapling six double-sided sheets of paper. Then the pages of the booklet are numbered in order from 1 through 24. Page 16 is on the same sheet of paper as what other 3 page numbers? (2006 MathIsCool)

3. The 16 squares on a piece of paper are numbered as shown in the diagram. While lying on a table, the paper is folded in half four times in the following sequence:

    (1) fold the top half over the bottom half

    (2) fold the bottom half over the top half

    (3) fold the right half over the left half

    (4) fold the left half over the right half.

    Which numbered square is on top after step 4? (1991 AMC8)

1	2	3	4
5	6	7	8
9	10	11	12
13	14	15	16

4. Each corner of a rectangular prism is cut off. Two (of the eight) cuts are shown. How many edges does the new figure have? Assume that the planes cutting the prism do not intersect anywhere in or on the prism. (1990 AMC8)

5. The figure may be folded along the lines shown to form a number cube. Three number faces come together at each corner of the cube. What is the largest sum of three numbers whose faces come together at a corner? (1989 AMC8)

	1		
6	2	4	5
	3		

6. Suppose a square piece of paper is folded in half vertically. The folded paper is then cut in half along the dashed line. Three rectangles are formed-a large one and two small ones. What is the ratio of the perimeter of one of the small rectangles to the perimeter of the large rectangle? (1989 AMC8)

7. The square in the first diagram "rolls" clockwise around the fixed regular hexagon until it reaches the bottom. In which position will the solid triangle be in diagram 4? (1988 AMC8)

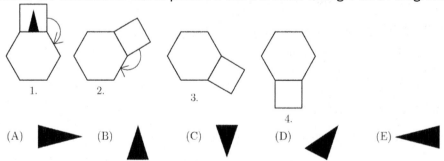

Pattern Recognition

[Basics] number, sense of spatial thinking, triangular numbers

Start from a Poke Magic

To have some more "whole-brained" exercise, let me introduce a poke magic using rotational symmetry concept. If an object can find its match by rotating itself by less than 360 degrees around a certain center point, we say this object has rotational symmetry. Problems of rotational symmetry are very good to train our spatial thinking skill. Here is a basic one: "how many letters from the 26 English alphabet have rotational symmetry?" Pay attention to the fact that the letter "d" is not qualified for a rotational symmetry, although it matches letter "p" after turning around for 180 degrees. The letter needs to match itself after rotation. So, the answer is five letters - "l", "o", "s", "x", and "z".

We can apply the rotational symmetry to a poker card magic design. As a demo, you draw the following cards from a standard deck.

Spade: ♠K, ♠Q, ♠J, ♠10, ♠4, ♠2
Heart: ♥ K, ♥ Q, ♥ J, ♥ 10, ♥ 4, ♥ 2
Club: ♣ K, ♣ Q, ♣ J, ♣ 10, ♣ 4, ♣ 2
Diamond: ♦K, ♦Q, ♦J, ♦10, ♦9, ♦8, ♦6, ♦5, ♦4, ♦3, ♦2, ♦1

What have you found that these cards have in common? They all have an attribute of rotational symmetry. If you rotate any of these cards 180 degree, it will look like the same as it is at the original position. For example, the 4 of clubs and the 10 of diamonds are such kind of cards.

45

However, what we are going to take advantage of are the remaining cards in the deck. They don't have the attribute of rotational symmetry. That means after you rotate one of the cards 180 degrees, it will present a different design on the card from the original one before the rotation took place. For example, the 7 of spades is one of such type of card.

Here comes with a procedure of a magic game by using the cards without the attribute of rotational symmetry.

<u>Step-1</u>

In a card stack, you will only use these cards without the attribute of rotational symmetry. Usually you can find 52-30=22 cards with such design from a standard deck. You organize them in such a fashion that you can easily remember what is facing up or down on each card. For example, since the symbols on each of these cards are either facing up or facing down, you may put all of cards in a way that the number of the upward symbols on the card is more than the number of the downward symbols on it. To understand this, you can compare the following spade 7 with the one above. The 7 of spades has been rotated by 180 degree from the above one. Can you now tell the difference between them?

<u>Step-2</u>

Let your guest pick one card from the stack. Make sure you cannot see any of these cards.

<u>Step-3</u>

You replace the card back in the stack after rotating it 180 degrees. Be careful not to let your guest notice this rotation.

<u>Step-4</u>

You shuffle the cards and mix them up in a more random order in the stack, but don't let any card rotate 180 degrees.

<u>Step-5</u>

You check the cards and find out the one that has a different style of design on it. There must be only one card that has more downward symbols than upward ones on it, because originally you had arranged all of the cards with more upward symbols than downward ones on them and you have only "changed" one card's design by rotating it 180 degrees.

Pattern recognition skill is very important. It requires a well-trained instinct to recognize relations among numbers, pictures, as well as their relative positions, based on given information.

[Example]
$1 - 2 - 3 + 4 + 5 - 6 - 7 + 8 + 9 - 10 - 11 + \ldots + 1992 + 1993 - 1994 - 1995 + 1996 = ?$
(1996 AMC8)

[Solution]
Group every four numbers, like 1 - 2 − 3 + 4 = 0, you will see the total sum is 0.

[Example]
$$2\left(1 - \frac{1}{2}\right) + 3\left(1 - \frac{1}{3}\right) + 4\left(1 - \frac{1}{4}\right) + \cdots + 10\left(1 - \frac{1}{10}\right) =$$
(A) 45   (B) 49   (C) 50   (D) 54   (E) 55
(1998 AMC8)

[Solution]
1 + 2 + 3 + ... + 9 = 45. (A) is correct.

[Example]
A "stair-step" figure is made of alternating black and white squares in each row. Rows 1 through 4 are shown. All rows begin and end with a white square. The number of black squares in the 37th row is? (1985 AMC8)

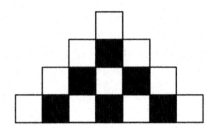

[Solution]
Note the first row had no black square, the number of black squares in any row is one less than the row number. The answer is 37 − 1 = 36.

[Example]
In the pattern below, the cat (denoted as a cat in the figures below) moves clockwise through the four squares, and the mouse (denoted as a mouse in the figures below) moves counterclockwise through the eight exterior segments of the four squares. (2003 AMC8)

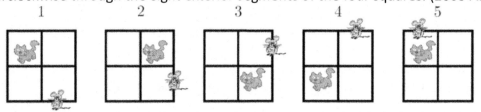

If the pattern is continued, where would the cat and mouse be after the 247th move?

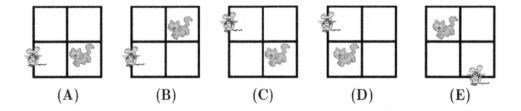

[Solution]
The cat has four possible positions that are repeated every four moves. 247 has a remainder of 3, when divided by 4. This corresponds to the position the cat has after the 3rd move, which is the bottom right corner.

The mouse has eight possible positions that repeat every eight moves. 247 has a remainder of 7, when divided by 8. This corresponds to the position the cat has after the 7th move, which is at the bottom edge on the left side of the grid.

Answer key is (A).

[Example]

If this path is to continue in the same pattern:

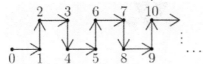

Then, which sequence of arrows goes from point 425 to point 427? (1994 AMC8)

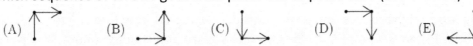

[Solution]

Notice the pattern from 0 to 4, 4 to 8, 8 to 12, ... repeats for every four arrows.

The remainder when 425 is divided by 4 is 1. The arrows from point 425 to point 427 correspond to points 1 and 3. Pattern (A) is a correct answer.

[Practice Problems]

1. A sequence of squares is made of identical square tiles. The edge of each square is one tile length longer than the edge of the previous square. The first three squares are shown. How many more tiles does the seventh square require than the sixth? (2002 AMC8)

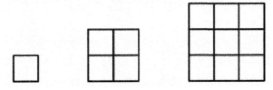

2. Terri produces a sequence of positive integers by following three rules. She starts with a positive integer, then applies the appropriate rule to the result, and continues in this fashion.

   Rule 1: If the integer is less than 10, multiply it by 9.

   Rule 2: If the integer is even and greater than 9, divide it by 2.

   Rule 3: If the integer is odd and greater than 9, subtract 5 from it.

   A sample sequence: 23, 18, 9, 81, 76, ...

   Find the 98th term of the sequence that begins 98, 49, ... (1998 AMC8)

3. If the pattern in the diagram continues, what fraction of the interior would be shaded in the eighth triangle? (1998 AMC8)

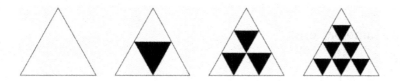

4. A rectangular board of 8 columns has squared numbered beginning in the upper left corner and moving left to right so row one is numbered 1 through 8, row two is 9 through 16, and so on. A student shades square 1, then skips one square and shades square 3, skips two squares and shades square 6, skips 3 squares and shades square 10, and continues in this way until there is at least one shaded square in each column. What is the number of the shaded square that first achieves this result? (1998 AMC8)

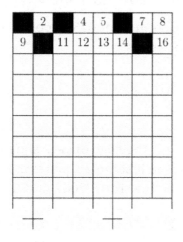

5. (1901 + 1902 + 1903 + ... + 1993) − (101 + 102 + 103 + ... + 193) = ? (1993 AMC8)

6. What number is directly above 142 in this array of numbers? (1993 AMC8)

$$
\begin{array}{ccccc}
& & 1 & & \\
& 2 & 3 & 4 & \\
5 & 6 & 7 & 8 & 9 \\
10 & 11 & 12 & \cdots &
\end{array}
$$

7. What is the 1992nd letter in this sequence? (1992 AMC8)

ABCDEDCBAABCDEDCBAABCDEDCBAABCDEDC ···

8. Which pattern of identical squares could NOT be folded along the lines shown to form a cube? (1992 AMC8)

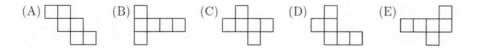

9. How many zeros are at the end of the product:
$25 \times 25 \times 25 \times 25 \times 25 \times 25 \times 25 \times 8 \times 8 \times 8$? (1991 AMC8)

10. An equilateral triangle is originally painted black. Each time the triangle is changed, the middle fourth of each black triangle turns white. After five changes, what fractional part of the original area of the black triangle remains black? (1991 AMC8)

 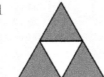 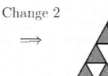

11. A list of 8 numbers is formed by beginning with two given numbers. Each new number in the list is the product of the two previous numbers. Find the first number if the last three are shown: ___?___ , _____ , _____ , _____ , _____ , _16_ , _64_ , _1024_
(1990 AMC8)

12. How many different patterns can be made by shading exactly two of the nine squares? Patterns that can be matched by flips and/or turns are not considered different. For example, the patterns shown below are **not** considered different. (1990 AMC8)

13. The letters A, J, H, S, M, E and the digits 1, 9, 8, 9 are "cycled" separately as follows and put together in a numbered list:

	AJHSME	1989
1.	JHSMEA	9891
2.	HSMEAJ	8919
3.	SMEAJH	9198

....

What is the number of the line on which "AJHSME 1989" will appear for the first time? (1989 AMC 8G/22)

14. Don is building a staircase pattern as shown in the figure. Each block is one foot high. How many blocks would it take to build steps that would be 20 feet high? (2012 MathIsCool)

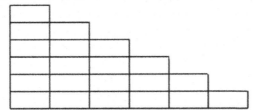

Three Cats

[Basics] ratio, proportion

This is a math problem dating back to ancient times. It asks, "If we know three cats can catch three mice in three minutes, how many such cats are wanted to catch 100 mice in 100 minutes?"

100 cats? Please take a breath and rethink it. Three cats use one minute to catch one mouse, so three cats will use 100 minutes to catch 100 mice. Seems easy, but is this correct?

The above answer is based on an assumption: the three cats work as a team to catch one mouse in one minute. However, unless it is clearly defined in the original question, there is other possible situations. What if three cats catch three mice independently? In this case, you may assume that one cat needs three minutes to catch one mouse. Consequently, one cat needs six minutes to catch two mice, nine minutes to catch three mice, ..., 99 minutes to catch 33 mice. Therefore, three cats need 99 minutes to catch 99 mice. Of course, this assumes each cat has the same productivity.

Can the three cats catch the 100th mouse within the 100th minute with or without working together? It seems the 4th cat is necessary in order to accomplish this job. To solve this problem correctly, we will need to make assumptions and the answer differs under different assumption.

[Example]
If it takes two chickens three days to lay five eggs, how many eggs will nine chickens lay in fourteen days? (2007 MathIsCool)

[Solution]
(9 x 14) x 5 / (2 x 3) = 105 eggs.

Chain of Ratios

[Example]
If eight "aarghs" are worth six "blahs", and ten "blahs" are worth three "crikeys", how many "crikeys" would 120 "aarghs" be worth? (2007 MathIsCool)

[Solution]
Build relation between (aarghs) and (crikeys) using (blahs) as a "bridge".
8 (aarghs) = 6 (blahs) ➜ 40 (aarghs) = 30 (blahs)

10 (blahs) = 3 (crikeys) ➔ 30 (blahs) = 9 (crikeys)
Then, 40 (aarghs) = 9 (crikeys) ➔ 120 (aarghs) = 27 (crikeys)

[Practice Problems]

1. Madam Hippo has her own currency consisting of Hips, Haps, and Hops. 28 Hips is the same as 16 Haps. 21 Haps is the same as 7 Hups. 14 Hups is the same as 4 Hops. If one United States dollar is equal to 7/8 of one Hip, how much, in US dollars, is one Hop worth? (2009 MathIsCool)

2. 24 Twinks is the same as 18 Zinks. 15 Zinks is the same as 3 Finks. 6 Finks is the same as 8 Blinks. How many Twinks are in a Blink? (2006 MathIsCool)

3. There are 3 widgets in a whatsit, 2 whatsits in 3 whoozits, 2 whoozits in 12 whatchamacallits, and 4 whatchamacallits in 5 wazzas. If Tom has 24 widgets, how many wazzas does he have? (2005 MathIsCool)

4. If there are two boinks in a bank, three banks in a bunch, and 5 bunches in a bin, how many bunches can I make from 2 binks and 6 boinks? (2001 MathIsCool)

5. Mort the Alien eats 1,500 solar systems every day. If there is 1 star per solar system, 8 planets for every star, and 17 moons for each planet, how many moons, planets, and stars does Mort eat in 2 days? (2016 MathIsCool)

6. Ton has 5 mangos. One mango can be traded for 4 durians, and each durian can be traded for either 3 apricots, or 2 peaches, or 1 orange. How many smoothies can Ton make, if each smoothie requires 1 durian, 2 apricots, 1 peach, and 1 orange? (2013 MathIsCool)

7. A number of students from Fibonacci Middle School are taking part in a community service project. The ratio of 8th graders to 6th graders is 5:3, and the ratio of 8th graders to 7th graders is 8:5. What is the smallest number of students that could be participating in the project? (2013 AMC8)

8. A recipe that makes 5 servings of hot chocolate requires 2 squares of chocolate, ¼ cup sugar, 1 cup water and 4 cups milk. Jordan has 5 squares of chocolate, 2 cups of sugar, lots of water and 7 cups of milk. If she maintains the same ratio of ingredients, what is the greatest number of servings of hot chocolate she can make? (2009 AMC8)

9. In a far-off land three fish can be traded for two loaves of bread and a loaf of bread can be traded for four bags of rice. How many bags of rice is one fish worth? (1999 AMC8)

Count

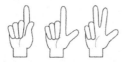

[Basics] number series, combination, and permutation

Everyone needs to learn how to count strategically. We learned to count numbers naturally from 1, 2, ..., all the way up to any big numbers that we are comfortable to count.

Number Series

[Example]
How do you count numbers in this series, 3, 4, 5, 6, ..., 100?

[Solution]
A common method is to convert the number series to something more straightforward. If we subtract 2 from every number in the series, we get 1, 2, 3, 4, ..., 98. We can tell immediately the count is 98, because we are so familiar with counting from one to any big integer number. Note the important thing here is, subtracting 2 from the every number in the original number series does not change the total count, but it gives us a much easier view of the number series, in terms of total count.

[Example]
What about 5, 8, 11, 14, ..., 101, then?

[Solution]
It looks more complicated than the previous one, but you can try the same approach,
Step-1. Subtract 5 from every number, it becomes 0, 3, 6, 9, ..., 96
Step-2. Divided by 3, it then becomes 0, 1, 2, 3, ..., 32
Step-3. It is not hard to count from 0, one by one up to 32. The total count is 33.
Step-4. The general term for i-th number in the series will be $x(i) = 3 * i + 5$, (i=0, 1, 2, ..., 32)
We are able to get not only the total count, but also a basic term for each number in the series.

Brute Force

Sometimes it is simpler to just list out all the possible items, when we count stuff. A "brutal" name to describe this approach is "Brute force".

[Example]

Bag A has three chips labeled 1, 3, and 5. Bag B has three chips labeled 2, 4, and 6. If one chip is drawn from each bag, how many different values are possible for the sum of the two numbers on the chips? (2011 AMC8)

[Solution]
By adding a number from Bag A and a number from Bag B together, the values we can get are 3, 5, 7, 5, 7, 9, 7, 9, 11. Therefore the number of different values is 5.

[Example]
My 12-hour digital clock shows hours and minutes, but not seconds. The sum of the digits of the time showing on my clock now is 3. In 165 minutes, the sum of the digits will be 9. What time is it now? (2010 MathIsCool)

[Solution]
Since the sum of the digits of the time showing on my clock now is 3, there are 10 different possible time now - 0:03, 0:30, 0:12, 0:21, 1:02, 1:20, 1:11, 2:01, 2:10, 3:00.
Adding 2:45 (equivalent to 165 minutes) to these time number (don't forget 2:66 should be corrected to 3:06 on a digital clock), you may quickly find 1:20 is the right answer. And, it could be 1:20 A.M. or 1:20 P.M., because this is a 12-hour digital clock.

[Example]

In the arrangement of letters and numerals below, by how many different paths can one spell AMC8? Beginning at the A in the middle, a path allows only moves from one letter to an adjacent (above, below, left, or right, but not diagonal) letter. One example of such a path is traced in the picture. (2017 AMC 8)

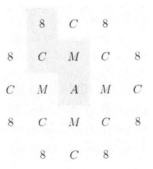

[Solution]

Looking at the connections between A, M, C, and 8, you find one A has four Ms next to it. One M links to three Cs. One C connects to two 8s. Therefore, in total, there are 4 x 3 x 2 = 24 different paths for "AMC8".

Rearrange Letters in a Word

When you rearrange letters in a word to form different words, an effective way is using basic combination or permutation concept. The key difference between these two methods is, combination does not care about order, while order does matter in permutation.

[Example]
How many different ways can you arrange the letters in the word "MATH"?

[Solution]
This is the same thing as "how do you arrange four people on the line?", which is a pure permutation problem. The answer is 4! = 24.

[Example]
How many ways can you rearrange the letters in SMALL?

[Solution]
This is equivalent to reorder five letters, among which there are two same letter "L". A simple mix of permutation with combination method is applicable. The answer is (5!) / (2!) = 60.

Pick Socks

[Example]
Claudine has 12 red socks, 5 purple socks, 5 white socks and 3 orange socks in a drawer. How many single socks must she draw at random to ensure a matching pair of socks? (1999 MathIsCool)

[Solution]
The worst case scenario is Claudine has picked four different color of socks first, then the fifth one will come up with a sock, which has the same color as one of the first four socks. Therefore, she must draw five socks at random to ensure a matching pair of socks among them.

[Example]
A five-legged Martian has a drawer full of socks, each of which is red, white or blue, and there are at least five socks of each color. The Martian pulls out one sock at a time without looking. How many socks must the Martian remove from the drawer to be certain there will be 5 socks of the same color? (2005 AMC8)

[Solution]

We always think about the edge case to solve this type of problem. The Martian can remove up to 12 socks, 4 of each color, before the 13th sock from any of the three color, at what time the Martian can be certain there will be 5 socks of the same color.

Select People

There are five people, A, B, C, D, and E.

Situation 1: When we want to pick two persons to play two different roles in a play.

Note AB and BA are different, because A and B are playing two different roles. We use permutation concept to figure out total number of different ways. We first pick one out of the five people to fill in the 1st role. There are five different ways to pick. Then, we pick one person from the remaining four persons to fill in the 2nd role. There are 5 – 1 = 4 different ways to pick now. Therefore, in total there are 5 x 4 = 20 different combinations.

Situation 2: When we want to pick two persons to attend a meeting.

Note AB and BA are now no longer making any difference in attending a meeting, according to the context. In this situation, we use pure combination to figure out total number of different ways, which is (5!) / (2! 3!) = 10.

Situation 3: When we want to pick two persons to watch a baseball game.
We need to be careful about the context here, whether the seats or tickets matter?
If the answer is yes, then this is the same one as situation 1. Or, it is situation 2.

Situation 4: When we want to divide five persons into two groups. One group has three persons. The other has two persons.

This is essentially a similar thing as situation 2. We pick two persons to form the first group, without concerning roles. The left three persons naturally form the second group. The count is 5! / (2! X 3!) = 10.

What if there are five balls, two blue and three green?
Since we usually assume two blue balls are identical and three green balls are indistinct,
   (a) Picking three balls (three blue, or two blue and one green, or one blue and two green, or three green) out of five is a combination, not a permutation problem.
   (b) Lining up the five balls is the same case as situation 4 above.

[Example]
Three friends have a total of 6 identical pencils, and each one has at least one pencil. In how many ways can this happen? (2004 AMC8)

[Solution]
For each person to have at least one pencil, assign one of the pencil to each of the three friends so that you have 3 left. In partitioning the remaining 3 pencils into 3 distinct groups, use "ball-and-urn" method to find the number of possibilities is a combination for "(3-1) out of (3 + 3 − 1)", which equals to 10.

<u>Note</u>
The "ball-and-urn" mathematical method is used to solve problems of such type as "how many ways can one distribute k indistinguishable objects into n bins?"
You may think about this as finding the number of ways to drop k balls into n urns. This is equivalent scenario as dropping k balls amongst n − 1 dividers.

The number of ways to accomplish the scenario is $\binom{n + k - 1}{k}$.

[Practice Problems]
1. Anna has four M&Ms (two red, one yellow, and one blue), which are identical except for color. In how many distinct (different) orders can Anna eat her M&Ms if each mouthful is either a single M&M or two M&Ms? (2009 MathIsCool)

2. How many different ways can you rearrange the letters in the word "MOON"?

3. Cory has 5 pairs of Nike shoes, 4 pairs of Adidas shoes, and 6 pairs of Puma shoes. If she picks shoes without looking, how many shoes does she need to pick in order to be sure that she has one right and one left shoe of the same brand? (2008 MathIsCool)

4. A gumball machine contains 9 red, 7 white, and 8 blue gumballs. What is the least number of gumballs a person must buy to be sure of getting four gumballs of the same color? (1994 AMC8)

5. An ATM password at Fred's Bank is composed of four digits from 0 to 9, with repeated digits allowable. If no password may begin with the sequence 9, 1, 1, then how many passwords are possible? (2016 AMC8)

6. Rebecca goes to the store where she buys five plants. If the store sells three types of plants, how many different combinations of plants can she buy? (2005 MathIsCool)

7. For her birthday, Clara wanted a balloon bouquet. If there are four colors to choose from and her bouquet will have four balloons, how many combinations of colored balloons are possible? (2005 MathIsCool)

8. Five dogs (Fido, Ruff, Lassie, Odie, and Snoopy) were playing in the park. Altogether, they had 5 identical Frisbees. In how many ways could the 5 Frisbees be distributed among the dogs when they leave the park to go home? (2010 MathIsCool)

9. Doris loves making outfits. An outfit consists of a matching pair of shoes, a hat, a shirt, and a skirt. Doris has 3 hats, 5 shirts, 4 skirts, and 14 pairs of shoes, including exactly one yellow hat, one yellow shirt, and one yellow skirt (but no yellow shoes). If she never wears two or more yellow items at the same time, how many different outfits can Doris make? (2008 MathIsCool)

10. Andy, Bob, and Colin are waiting in line at the Googolplex Theater to see the new Star Wars movie. Andy is the 14th person in line, Bob is the 78th, and Colin is further back. There are half as many people standing between Andy and Bob as between Bob and Colin. What number person in line is Colin? (2006 MathIsCool)

11. How many positive three-digit numbers contain exactly two distinct digits (e.g. 343 or 772, but not 589 or 111)? (2006 MathIsCool)

12. How many positive two-digit numbers contain at least one digit of 1 but not the digit of 2? (2006 MathIsCool)

13. How many numbers between 10 and 50 (inclusive) are divisible by 2, 3 or 5? (2006 MathIsCool)

14. Soda is sold in packs of 6, 12 and 24 cans. What is the minimum number of packs needed to buy exactly 90 cans of soda? (2005 AMC8)

15. Two-thirds of the people in a room are seated in three-fourths of the chairs. The rest of the people are standing. If there are 6 empty chairs, how many people are in the room? (2004 AMC8)

16. At a party there are only single women and married men with their wives. The probability that a randomly selected woman is single is 2/5. What fraction of the people in the room are married men? (2004 AMC8)

17. Stacey has 1000 sticks. She groups them into bundles of 6, and when she gets 6 such bundles, she ties them together to form a bindle. When she gets 6 bindles, she ties them together to form a bandle. When she has finished, Stacey has A bandles, I bindles that are not in bandles, U bundles that are not in bindles, and S sticks that are not in bundles. Write these four values in order (AIUS). (2010 MathIsCool)

18. A token exchange machine takes an exact number of tokens of one color and returns an exact number of tokens of another color. Five purple tokens can be exchanged for 3 green tokens, or 4 purple tokens can be exchanged for 3 yellow tokens. Five green tokens can be exchanged for 2 red tokens. In each case, the reverse exchange is also possible (eg, 3 yellow tokens can be exchanged for 4 purple tokens). If I start with 9 red tokens and get as many yellow tokens as possible, how many tokens in all will I have when I finish? (2010 MathIsCool)

19. A box contains gold coins. If the coins are equally divided among six people, four coins are left over. If the coins are equally divided among five people, three coins are left over. If the box holds the smallest number of coins that meets these two conditions, how many coins are left when equally divided among seven people? (2006 AMC8)

Count with Venn diagram

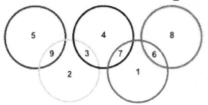

[Basics] counting, number operations

[Example]
Mathville High has 50 students enrolled and every student is taking at least one math or science class. 25 of the students take only a math class while 7 take both math and science. How many students take only science?

[Solution]

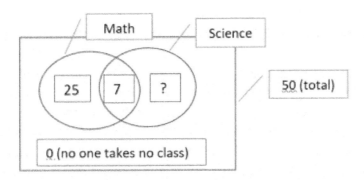

Once we draw a simple Venn diagram with all the information presented in it, the answer becomes as clear as, 50 - 0 – 25 – 7 = 18 students, who takes only science class.

[Example]
Of the 64 proctors volunteering today, all are on the Mt. Spokane Math Team. 36 are taking AP Calculus, 18 are taking AP Biology, 16 are taking AP English, 4 are taking AP Biology and AP Calculus, 7 are taking AP Biology and AP English and 5 are taking AP Calculus and AP English. Seven are not taking any AP courses. How many are taking all three courses: AP Biology, AP Calculus and AP English at the same time? (2006 MathIsCool)

[Solution]
We first subtract the number of proctors not taking any of the three AP classes from the total number of proctors (so we can focus on the proctors taking AP classes): 64 – 7 = 57.
Looking at the Venn. Diagram, we can assume there are x proctors taking all three courses AP Biology, Calculus, and English, then,
- Taking both AP Biology and AP Calculus, but not AP English ➔ 4 - x
- Taking both AP English and AP Biology, but not AP Calculus ➔ 7 - x
- Taking both AP Calculus and AP English, but no AP Biology ➔ 5 – x

And then,
- Taking only AP Biology ➔ $18 - (7 - x + x + 4 - x) = 7 + x$
- Taking only AP English ➔ $16 - (7 - x + x + 5 - x) = 4 + x$

Because,

$$(4 + x) + (7 - x) + (7 + x) + 36 = 57 \ ➔ \ 18 + x = 21 \ ➔ \ x = 3$$

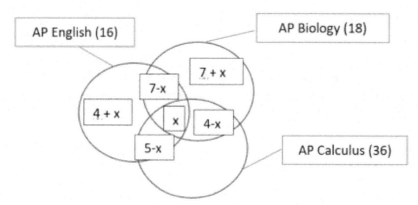

If you are not used to the equation with x variable, you may refer to the following explanation.

We first add up the total number of students in AP Calculus, AP Biology, and AP English: $36 + 18 + 16 = 70$. However, we have counted the students in two or more of these classes twice, so we must subtract the number of students taking two or more of these classes at the same time once: $70 - 4 - 7 - 5 = 54$. However, now we have added the number of students taking all three classes at the same time three times (in the 36, 18, and 16), but we have also subtracted this number of students three times (in 4, 7, 5). Since these students are still counted in the total proctor count who take AP classes, we must add this number of students back. Earlier we have got total number of students taken any AP class is 57. This number is $57 - 54 = 3$.

[Example]
Mrs. Sanders has three grandchildren, who call her regularly. One calls her every three days, one calls her every four days, and one calls her every five days. All three called her on December 31, 2016. On how many days during the next year did she not receive a phone call from any of her grandchildren? (2017 AMC 8)

[Solution]
Drawing a Venn diagram, you will find

- the number of days that each grandchild called Mrs. Sanders is $\left\lfloor \dfrac{365}{3} \right\rfloor = 121$, $\left\lfloor \dfrac{365}{4} \right\rfloor = 91$, $\left\lfloor \dfrac{365}{5} \right\rfloor = 73$

- the number of days that two of the three grandchildren called Mrs. Sanders is $\left\lfloor \dfrac{365}{12} \right\rfloor = 30$, $\left\lfloor \dfrac{365}{15} \right\rfloor = 24$, $\left\lfloor \dfrac{365}{20} \right\rfloor = 18$

- the number of days that three of them called Mrs. Sanders is $\left\lfloor \frac{365}{60} \right\rfloor = 6$

The total number of days that Mrs. Sanders received at least one call is:

$(121 + 91 + 73) - (30 + 24 + 18) + 6 = 285 - 72 + 6 = 219$

Thus, the number of days that she didn't receive any call from her grandchildren is:

$365 - 219 = 146$

[Summary]
Using Venn diagram is a visualized approach. It effectively breaks down the whole piece into multiple areas according to definition, and count number in each area. Finally it adds or subtracts relevant numbers and gets the answer key.

[Practice Problems]

1. Sets A and B, shown in the Venn diagram, have the same number of elements. Their union has 2007 elements and their intersection has 1001 elements. Find the number of elements in A. (2007 AMC8)

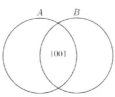

2. Mr. Sutter has an interesting class. 20 of his students can speak English and 10 of his students can speak Spanish. 5 of his students can speak both Spanish and English. How many students are in his class? (1999 MathIsCool)

3. In a town of 351 adults, every adult owns a car, motorcycle, or both. If 331 adults own cars and 45 adults own motorcycles, how many of the car owners do not own a motorcycle? (2011 AMC8)

4. Each of the 39 students in the eighth grade at Lincoln Middle School has one dog or one cat or both a dog and a cat. Twenty students have a dog and 26 students have a cat. How many students have both a dog and a cat? (2008 AMC8)

5. Mathland Middle School's math team members all participate in at least one of three other school activities: band, leadership, and track. Of the 40 members of the team, 15 are in leadership, 23 are on the track team, 4 are in both leadership and band, 6 are in both leadership and track, 3 are in band and track, and 2 students are in all three activities. How many students are in the band, total? (2005 MathIsCool)

6. At Mt. Rainier High School "D" students are on the debate team, "C" students are on the chess team, and "M" students are on the math team. "X" students are on the math team and the debate team, "Y" students are on the chess team and the debate team, and "Z" students are on the math team and chess team. "A" students are on all three teams. In terms of M, X, Z, and A, how many students are only on the math team? (2004 MathIsCool)

7. Thirty-seven members of the Lewis and Clark Math team were told to write the 6th grade "Math is Cool" contest. Fourteen members wrote the contest, fifteen members typed the contest, fifteen members proofread the contest, five members typed and proofread, seven members wrote and proofread, and four members wrote and typed the contest. If six members goofed off and did nothing, how many participated in all three activities? (2004 MathIsCool)

8. The Pythagoras High School band has 100 female and 80 male members. The Pythagoras High School orchestra has 80 female and 100 male members. There are 60 females who are members in both band and orchestra. Altogether, there are 230 students who are in either band or orchestra or both. What is the number of males in the band who are NOT in the orchestra? (1991 AMC8)

9. In a room, 2/5 of the people are wearing gloves, and ¾ of the people are wearing hats. What is the minimum number of people in the room wearing both a hat and a glove? (2010 AMC8)

Chessboard

[Basics] count

Imagine that you are playing chess games with two Grand Masters. How can you find a way to win one of them, or draw both of them? The solution to this dilemma can be found in a movie called "Tomorrow Never Dies"; you may be inspired!

[Question]
A long, long time ago in a kingdom, a King asked one man what he wanted the most. The man said, "My majesty, could you have someone put one grain (a single grain of wheat) on the first square, two grains on the second square, four grains on the third, so on (twice as many grains as on the previous square), until the last square on the chess board." The king ordered someone to do it immediately because he thought this was too easy. However, the King soon realized he and his kingdom would run into a big trouble. Do you know what trouble was it?

[Answer]
There are 8 x 8 = 64 squares on the squared chessboard. In the universe of numbers, 64 is one of the amazingly-magical ones. You may be surprised to find out that the last square was to be placed with grains 2^{63}, which is a huge number. It was big enough to bankrupt the Kingdom.

[Example]
The numbers are presented on the squares on a chessboard. Once it reaches the last square (hH), it will continue with the next number at the first square (aA). Can you figure out on which square the number "2005" will appear?

	A	B	C	D	E	F	G	H
a	1	2	3	4	5	6	7	8
b	9	10	11	12	13	14	15	16
c	17	18	......					
d								
e								
f								
g								
h								

[Solution]
Mod(2005, 64) = 21
Mod(21, 8) = 5
[21/8] = 2 ➔ cE

[Example]
How many 7 x 7 squares on the 8 x 8 chessboard? How many 6 x 6 squares?

[Solution]
No. of 7 x 7 squares ➔ 2 x 2 = 4
No. of 6 x 6 squares ➔ 3 x 3 = 9

[Example]
How many ways to put 8 mutually non-attacking rooks on a standard chessboard?

[Solution]
This is a permutation counting problem indeed. The answer is 8!.

[Practice Problems]

1. One pebble is put on one square of a chessboard. Then four pebbles are put on another square of the chessboard. Again, seven pebbles are put on another square of the chessboard. Again ten pebbles are put on another different square of the chess board. This process continues until all 64 squares of the chessboard contain pebbles. Find the total number of pebbles on the chessboard.

2. Which of the parts of a chessboard illustrated below cannot be covered exactly and completely by a whole number of non-overlapping dominoes? A "domino" is made up of two small squares:

   (A) 3 × 4   (B) 3 × 5   (C) 4 × 4   (D) 4 × 5   (E) 6 × 3

3. How many rectangles on an 8 x 8 chessboard?

4. The 64 squares on a chessboard are numbered as shown in the diagram. While lying on a table, the paper is folded in half four times in the following sequence:
   (1) fold the top half over the bottom half
   (2) fold the bottom half over the top half
   (3) fold the right half over the left half
   (4) fold the left half over the right half

   How many numbered squares are on top after step (4)? What are these numbers?

1	2	3	4	5	6	7	8
9	10	11	..	..			
			...		54	55	56
57	58	59	60	61	62	63	64

5. On a virtual chessboard (8 x N) that has 8 columns, row one is numbered 1 through 8, row two is 9 through 16, and so on. A student shades square 1, then skips one square and shades square 3, skips two squares and shades square 6, skips three squares and shades square 10, and continues in this way until there is at least one shaded square in each column. What is the number of the shaded square that first achieves this result? (1998 AMC8)

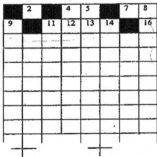

6. A beetle sits on each square of a 9 x 9 checkerboard. Each beetle can crawl diagonally to a neighboring square, leaving some squares empty and others with multiple beetles. What is the smallest possible number of empty squares after each beetle has moved exactly once? (2002 MathIsCool)

Coin, Die, and Card

[Basics] count, probability

"Fate laughs at probabilities and chance (probability) governs all." Many board games use coin, die, or card to demonstrate application of probability concept. "Probability" describes the possibility of what might happen, before it might happen. The term "what" implies an event, or a group of events.

Probability of one single event:
P = #(expected outcomes) / #(all possible outcomes)   (#: the count of)

Probability of a group of events:
P = Sum or Multiplication, of each single event's probability

How to count outcomes?
   Basic Rules:
      1. Each possible outcome should have equal chance to occur
      2. Impossible to have negative count of expected outcomes and all outcomes
      3. Impossible to have more count of expected outcomes than all possible outcomes.

   Use counting techniques to compute number of possible outcomes
      Permutation (the order is also counted)
      Combination

What are types of events in probability?
      1. Single event,
      2. Complement of an event,
      3. Independent events,
      4. Dependent event,
      5. Compound event

In this section, we will be looking into event type 1 and 3 in the following examples. By the way, event type 2 is the opposite of event type 1, conceptually. We will demonstrate how to work with event type 4 and type 5 in a later section.

Coin scenario:
      1. If I flip a coin, what is the probability to get the head?
          This is a single event.

2. If I flip a coin twice, what is the probability to get the head twice?
    This has two independent events as described in 1.

Die / Dice scenario:
1. If I roll a 6-sided die once, what is the probability to get the face with number "3"?
    This is a single event.
2. If I roll three 6-sided dice once for each, how many possible outcomes here?
    This has three independent events.

Is it described in scenario 1 the same thing as rolling one die for three times?
Yes, it is the same thing, if all faces on the three dice are evenly made.

Card scenario:
1. What is the probability of drawing a King card from a standard deck (52 cards)?
    This is a single event.
2. What is the probability of drawing a queen or a spade from a standard deck?
    This has two independent events.

A correct way to deal with probability problem is first identify the type of event and then select a right method to solve it.

[Example]
Mary has a standard deck of 52 playing cards. What are the chances that she will draw a red King or a diamond?

[Solution]
It is about probability of independent events.
Event to draw a red King: there are two red King cards out of total 52 cards, so its probability is 2/52.
Event to draw a diamond: there are 13 diamond cards out of total 52 cards, so its probability is 13/52.
Since it is about the "OR" relation between the above two events, the answer is 2/52 + 13/52 − 1/52 = 14/52 = 7/26, as we need to remove one double counted sample (i.e. diamond King), which has been counted twice by the two events.

[Example]
What is the probability of first drawing a red King, putting the card back into the deck, and then drawing a red Ace out of a standard deck containing 52 cards?

[Solution]
In this problem, there are two independent events with "AND" relation.
Event of drawing a red King: 2/52 = 1/26
Event of drawing a red Ace: 2/52 = 1/26.

Pay attention to the phrase – "…putting the card back into the deck, and then…", which indicates "AND" relation of the two events. Its probability is: (1/26) x (1/26) = 1/676.

[Example]
If two dice are tossed, what is the probability that the product of the numbers showing on the tops of the dice is greater than 10? (1992 AMC8)

[Solution]
When the first dice is 1, there is no chance. When the first dice is 2, the second dice can be 6. When the first dice is 3, the second dice can be 4, 5, 6. When the first dice is 4, the second dice can be 3, 4, 5, 6. When the first dice is 5, the second dice can be 3, 4, 5, 6. When the first dice is 6, the second dice can be 2, 3, 4, 5, 6. The total number of ways for the product to be greater than 10 is 1 + 3 + 4 + 4 + 5 = 17 and the total number of possibilities is 6 x 6 = 36, yielding a probability of 17/36.

[Example]
The two wheels shown below are spun and the two resulting numbers are added. What is the probability that the sum is even? (1994 AMC8)

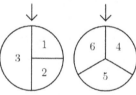

[Solution]
An even sum occurs when an even is added to an even or an odd is added to an odd. Looking at the areas of the regions, the chance of getting an even in the first wheel is ¼ and the chance of getting an odd is ¾. On the second wheel, the chance of getting an even is 2/3 and an odd is 1/3. (1/4) x (2/3) + (3/4) x (1/3) = 5/12.

[Practice Problems]

1. Find the probability that you will draw a 9 from a standard deck of 52 cards and at the same time flip 4 fair coins and get all heads.
    (A) 1/208   (B) 1/2   (C) 29/208   (D) 1/29

2. A fair coin is tossed 3 times. What is the probability of at least two consecutive heads? (2013 AMC8)

3. A fair 6-sided die is rolled twice. What is the probability that the first number that comes up is greater than or equal to the second number? (2011 AMC8)

4. Two cards are dealt from a deck of four red cards labeled A, B, C, D and four green cards labeled A, B, C, D. A winning pair is two of the same color or two of the same letter. What is the probability of drawing a winning pair? (2007 AMC8)

5. When a fair six-sided dice is tossed on a table top, the bottom face cannot be seen. What is the probability that the product of the faces that can be seen is divisible by 6? (2003 AMC8)

6. Harold tosses a coin four times. What is the probability that he gets at least as many heads as tails? (2002 AMC8)

7. Two dice are thrown. What is the probability that the product of the two numbers is a multiple of 5? (2001 AMC8)

8. Keiko tosses one penny and Ephraim tosses two pennies. What is the probability that Ephraim gets the same number of heads that Keiko gets? (2000 AMC8)

9. A complete cycle of a traffic light takes 60 seconds. During each cycle the light is green for 25 seconds, yellow for 5 seconds, and red for 30 seconds. At a randomly chosen time, what is the probability that the light will NOT be green? (1999 AMC8)

10. A pair of 8-sided dice have sides numbered 1 through 8. Each side has the same probability (chance) of landing face up. What is the probability that the product of the two numbers that land face-up exceeds 36? (1997 AMC8)

11. Diana and Apollo each roll a standard die obtaining a number at random from 1 to 6.. What is the probability that Diana's number is larger than Apollo's number? (1995 AMC8)

12. Each spinner is divided into 3 equal parts. The results obtained from spinning the two spinners are multiplied. What is the probability that this product is an even number? (1991 AMC8)

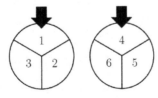

13. Every time these two wheels are spun, two numbers are selected by the pointers. What is the probability that the sum of the two selected numbers is even? (1989 AMC8)

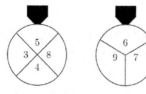

14. Ten balls numbered 1 to 10 are in a jar. Jack reaches into the jar and randomly removes one of the balls. Then Jill reaches into the jar and randomly removes a different ball. What is the probability that the sum of the two numbers on the balls removed is even?

15. Alexander has a box with full of toy cars in it. He counts all of the toy cars in the box and discovers that he has 9 blue cars, 3 red cars, 4 green cars, and 1 black car. Alexander randomly takes one toy car out of the box and then passes it to Amanda. He then picks up another toy car from the box and also passes it to Amanda. If a/b is the possibility that both of the toy cars that Alexander passed to Amanda were green (a, b are positive integers and a/b is a correctly reduced fraction), then what is a + b?

16. What is the probability of drawing two cards of the same suite (Spades, Hearts, Clubs, or Diamonds) from a standard deck of cards (52 in total)?

17. Two standard 6-sided dice are rolled. The probability of rolling a sum of x is 1/12. Find the product of all possible integers of x. (2004 MathIsCool)

18. At the Annual Math Open held at Cow Pi Beach, the Mathletes outnumbered the Trimathletes 7: x. If a competitor was chosen at random, the probability that it was a Trimathlete was y/12. What is the product of x and y?

19. An integer between 1000 and 9999, inclusive, is chosen at random. What is the probability that it is an odd integer whose digits are all distinct? (2017 AMC 8)

Game of Probability

[Basics] probability, geometric number series

Dependent Probability

When the probability outcome of one event doesn't affect the outcome of the other event, we say the two events are independent. For example, when you throw a die and pick a card from a deck, these two events are independent, because the card you choose has no dependency on the resulting face of a die. However, if you don't put the card back and choose a card again from the remaining deck, the outcome will be dependent on the outcome from your first pick. If your first card was a spade king, the probability you pick the same spade king on your second pick will be 0. If you didn't pick a spade king the first time, then the probability you pick a spade king next time will be 1/51, assuming you didn't put the first card back to the deck. As you can see, the outcome of the second event is dependent on the outcome of the first event.

[Example]
As a joke, Jake put 4 boiled eggs in a carton. The other 8 eggs in the carton were raw. Jake's mom took 2 eggs from the carton at random to make pancakes. At least one of these eggs was boiled. As a fraction, what is the probability that both eggs she took were boiled? (2010 MathIsCool)

[Solution]
P1 (at least one boiled eggs taken from 12 eggs) = 1 - C(8, 2) / C(12, 2) = 1 – 28/66 = 19/33
P2 (two boiled eggs taken from 12 eggs) = C(4, 2) / C(12, 2) = 6/66 = 1/11
P (two boiled eggs, given at least one boiled eggs) = P2 / P1 = (1/11) / (19/33) = 3/19.

[Example]
Freddy takes one ball at random from a bag with 4 blue, 7 red, and 3 green balls, and keeps it. Freddy can tell that this ball is either red or green, but he can't tell which because he is colorblind. Find the probability that the next ball Freddy takes from the bag at random will be red. (2009 MathIsCool)

[Solution]
There are two approaches.

Approach (a):

The probability of next ball will be red is dependent on the color of current ball taken.

P (next ball will be red, given current ball is either red or green from all balls) =

P (current ball is red from red and green balls) x (6/13) + P (current ball is green from red and green balls) x (7/13)

→ (7/10) (6/13) + (3/10) (7/13) = 63/130

Approach (b):

P1 (current ball is either red or green from all balls) = 10/14

P2 (next ball will be red) = (3/14) (7/13) + (7/14) (6/13)

P (next ball will be red, given current ball is either red or green from all balls) = P2/P1

→ (63 / (14 x 13)) / (10/14) = 63/130

Recursive Probability

Case-1: Two players

A and B in turns throw a coin, whoever gets head first, wins the game, given they have even chances (1/2 for each).

Every round A and B has the same ratio of winning probability.

1st round: P1(A win) = ½, P1(B win) = (1 − P1(A win)) x ½ = ¼ , ...

Overall P(A win) = P1(A win) + P2(A win) + ...

Overall P(B win) = P1(B win) + P2(B win) + ...

P(A win) / P(B win) = P1(A win) / P1(B win) = 2, and P(A win) + P(B win) = 1, therefore,

P(A win) = 2/3, P(B win) = 1/3.

Case-2: Three players

Using the same approach from Case-1, we may conclude the followings.

A, B, C in turns throw a coin, whoever gets head first, wins the game.

P(A win) = 4/7, P(B win) = 2/7, and P(C win) = 1/7

A, B, C in turns throw a die (six-sided), whoever gets "6", wins the game.

P(A win) = 36/91, P(B win) = 30/91, and P(C win) = 25/91

When we have four persons, A, B, C and D, their winning probability numbers have a pattern.

[Example]

Colin, Kai, and Sampson are playing a dice game. Colin goes first, Kai second, and Sampson third. The first person to roll a three wins and the game is over. The dice are continuously passed in the above order until someone wins. What is the probability that Sampson wins? (2008 MathIsCool)

[Solution]

At the first round, each person's winning probabilities are:

P1(Colin) = 1/6, P1(Kai) = (5/6) (1/6) = 5/36, P1(Sampson) = (5/6)(5/6)(1/6) = 25/216;
Since the overall P(Colin) : P(Kai) : P(Sampson) = P1(Colin) : P1(Kai) : P1(Sampson) = 36:30:25,
and P(Colin) + P(Kai) + P(Sampson) =1, P(Sampson) = 25/91.

We can also check by the following calculation.
At the second round, they become:
P2(Colin) = (5/6)^3 (1/6), P2(Kai) = (5/6)^3 (5/36), P2(Sampson) = (5/6)^3 (25/216);
......
The outcome of each person's winning probability forms a geometric number series.
The overall P (Sampson wins) = P1(Sampson) + P2(Sampson) + ... + Pn(Sampson) + ...
Following sum formation of geometric number series,
P (Sampson wins) = (25/216) (1 - (5/6)^3)^n / (1 - (5/6)^3) = (25/216) / (1 – 125/216) = 25 / 91.
n can be an unlimitedly big integer, as the game continues forever.
Same process applies to the other two persons, to find out their winning probabilities.

[Practice Problems]

1. A dog has three different toys, but plays with its favorite toy exactly twice as often as either
    of the other two. If the dog is currently playing with a toy, what is the likelihood that it is
    playing with its favorite toy? Give your answer as a fraction. (2008 MathIsCool)

2. There is a 1/3 chance that it will rain today. If it does rain, there is a ¾ chance that Sarah will
    stay home. If it does not rain, there is only a ¼ chance that she will stay home. What is the
    probability that Sarah stays home? (2017 MathIsCool)

3. Two cards are placed in a hat. One has two black sides and the other has a black side and an
    orange side. If Joel draws a card and sees that one side is black, what is the probability the
    other side is black? (2002 MathIsCool)

4. Jamie and Julie are on a roller coaster ride at Silverwood Park. The probability that Julie will
    let go of the bar is 1/3. When riding alone or together, Jamie will scream with a probability
    of 1/4. However, when they ride together, the probability that Jamie screams is doubled if
    Julie lets go of the bar. As a reduced fraction, what is the probability that Jamie screams
    when the two ride together? (2009 MathIsCool)

5. Biff, Eho and Frank are tossing a coin to see who can gets the first head. As soon as a head
    appears (even if each has not had an equal number of tosses) the game is over. If Biff tosses
    first, then Eho, then Frank, with the order repeated (possibly indefinitely) until the first
    head appears, what is the probability that Frank will win? (2007 MathIsCool)

6. Ozzie and Colin are playing a game with three dice where Colin rolls one and Ozzie rolls two.
    Colin wins if his roll equals or betters the sum of Ozzie's two dice (Ozzie likes this game).
    What is the probability Colin wins any given round? (2007 MathIsCool)

Count in Graph

[Example]
How many triangles are in this figure? (Some triangles may overlap other triangles.) (1998 AMC8)

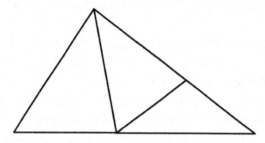

[Solution]
By inspection, we have 5 triangles: Each of the 3 small triangles, 1 medium triangle made of the rightmost two small triangles, and the 1 large triangle.

[Example]
Each of the twenty dots on the graph below represents one of Sarah's classmates. Classmates who are friends are connected with a line segment. For her birthday party, Sarah is inviting only the following: all of her friends and all of those classmates who are friends with at least one of her friends. How many classmates will not be invited to Sarah's party? (2003 AMC8)

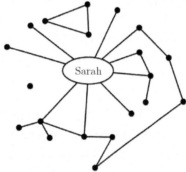

[Solution]
There are 3 people who are friends with only each other who won't be invited, plus 1 person who has no friends, and 2 people who are friends of friends of friends who won't be invited. So the answer is 6.

[Example]
Jane can walk any distance in half the time it takes Hector to walk the same distance. They set off in opposite directions around the outside of the 18-block area as shown. When they meet for the first time, they will be closest to A, or B, or C, or D, or E? (1995 AMC8)

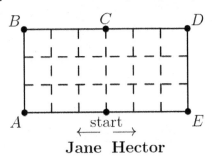

[Solution]
Counting around, when Jane walks 12 steps, she will be at D. When Hector walks 6 steps, he will also be at D. Since Jane has walked twice as many steps as Hector, they will reach this spot at the same time. Thus, the answer is D.

[Example]
Chuckles the elephant is trying to get to Paris! The map to Paris is shown below. If Chuckles can only walk along the lines going east or north, how many paths can she take? (2012 MathIsCool)

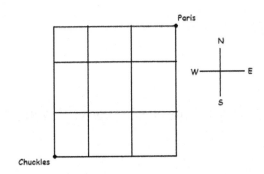

[Solution]
You could use combination or Pascal's triangle to figure out the answer, but the easiest way is illustrated in the following diagram. At every intersection, it is the sum of the count at its west and the count at its south. The intersection point at Paris has the count equal 10 + 10 = 20.

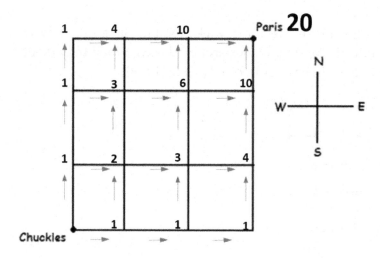

[Practice Problems]

1. A beetle is standing at point A. It only walks right and down and it always stays on the lines. How many different paths can it take to point B? (2006 MathIsCool)

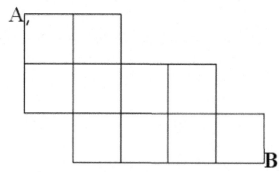

2. Toothpicks are used to make a grid that is 60 toothpicks long and 32 toothpicks wide. How many toothpicks are used altogether? (2013 AMC8)

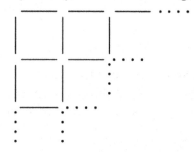

3. Three A's, three B's, and three C's are placed in the nine spaces so that each row and column contain one of each letter. If A is placed in the upper left corner, how many arrangements are possible? (2008 AMC8)

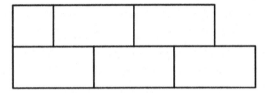

4. A block wall 100 feet long and 7 feet high will be constructed using blocks that are 1 foot high and either 2 feet long or 1 foot long (no blocks may be cut). The vertical joins in the blocks must be staggered as shown, and the wall must be even on the ends. What is the smallest number of blocks needed to build this wall? (2000 AMC8)

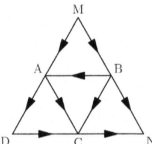

5. Using only the paths and the directions shown, how many different routes are there from M to N? (1986 AMC8)

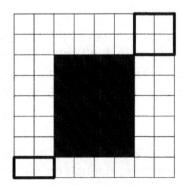

6. Samantha lives 2 blocks west and 1 block south of the southwest corner of City Park. Her school is 2 blocks east and 2 blocks north of the northeast corner of City Park. On school days, she bikes on streets to the southwest corner of City Park, then takes a diagonal path through the park to the northeast corner, and then bikes on streets to school. If her route is as short as possible, how many different routes can she take? (2013 AMC8)

Logical Reasoning

Making meaningful deductions from every single piece of information, including basic knowledge and the given data, is essential. If there is a conditional statement available in the given set of information, we can transform it to a "contrapositive" statement. The contrapositive statement will always be true, given the original statement is true.

What is a conditional statement? It has the "if a, then b" logical structure. To form the contrapositive of the conditional statement, we interchange the hypothesis and the conclusion of its inverse statement: "if NOT b, then NOT a". One example is, given a statement "If a triangle has one right angle in it, then the triangle is a right triangle." The contrapositive statement will be "If a triangle is not a right triangle, then the triangle has no right angle in it." We know both of these to be correct. The hypothesis and the conclusion in these two statements are interchangeable.

[Example]
Students guess that Norb's age is 24, 28, 30, 32, 36, 38, 41, 44, 47, and 49. Norb says, "At least half of you guessed too low, two of you are off by one, and my age is a prime number." How old is Norb? (2011 AMC8)

[Solution]
Given three pieces of key information, we can safely deduce step by step as below, until we reach conclusion.
Step 1. If at least half the guesses are too low, then his age must be greater than 36.
Step 2. If two of the guesses are off by one, then his age is in between two guesses whose difference is 2. It could 31, 37, or 48, but because it is greater than 36, it can only be 37 or 48.
Step 3. Lastly, his age is a prime number so the answer must be 37.

[Example]
Bridget, Cassie, and Hannah are discussing the results of their last math test. Hannah shows Bridget and Cassie her test, but Bridget and Cassie don't show theirs to anyone. Cassie says, 'I didn't get the lowest score in our class,' and Bridget adds, 'I didn't get the highest score.' What is the ranking of the three girls from highest to lowest? (2013 AMC8)

[Solution]
If Hannah did better than Cassie, there would be no way Cassie could know for sure that she didn't get the lowest score in the class. Therefore, Hannah did worse than Cassie. Similarly, if Hannah did worse than Bridget, there is no way Bridget could have known that she didn't get

80

the highest in the class. Therefore, Hannah did better than Bridget, so our order is Cassie, Hannah, and Bridget

[Example]
Amy, Bill and Celine are friends with different ages. Exactly one of the following statements is true.
     I.   Bill is the oldest.
     II.   Amy is not the oldest.
     III.   Celine is not the youngest.
Rank the friends from the oldest to the youngest. (2004 AMC8)

[Solution]
If Bill is the oldest, then Amy is not the oldest, and both statements I and II are true, so statement I is not true.
If Amy is not the oldest, and we know Bill cannot be the oldest, then Celine is the oldest. This would mean she is not the youngest, and both statements II and III are true, so statement II is not true.
Therefore, statement III is true, and both I and II are false. Amy is the oldest, Celine is in the middle, and Bill is the youngest. The order is Amy, Celine, Bill.

[Example]
Bo, Coe, Flo, Jo, and Moe have different amounts of money. Neither Jo nor Bo has as much money as Flo. Both Bo and Coe have more than Moe. Jo has more than Moe, but less than Bo. Who has the least amount of money? (1999 AMC8)

[Solution]
Analysis of each sentence in order to get clue:
1. "Neither Jo nor Bo has as much money as Flo." ➔ Flo clearly does not have the least amount of money. Eliminate Flo from the possible answer.
2. "Both Bo and Coe have more than Moe." ➔ Rule out Bo and Coe, since they clearly do not have the least amount of money.
3. "Jo has more than Moe". ➔ Rule out Jo.
Now the only person left is Moe, who should have the least amount of money.

[Example]
Five cards are lying on a table as shown.

81

Each card has a letter on one side and a whole number on the other side. Jane said, "If a vowel is on one side of any card, then an even number is on the other side." Mary showed Jane was wrong by turning over one card. Which card did Mary turn over? (1985 AMC8)

[Solution]
Logically, Jane's statement is equivalent to its contrapositive, "If an even number is not on one side of any card, then a vowel is not on the other side." This statement is also similar to "If an odd number is one side of any card, then a vowel is not on the other side". In order to show Jane wrong, Mary need to find a card with an odd number on one side, and a vowel on the other side. The only card that could possibly have this property is the card with "3" on it.

[Practice Problems]

1. Animals on the Hochstatter Farm:

	BROWN	BLACK
Goat	18	12
Cow	30	20
Horse	12	8

   Question A: Given that an animal is black on the Hochstatter farm, what is the probability the animal is a cow?
   Question B: What is the probability that an animal picked at random on the Hochstatter farm is a horse?
   Question C: What is the probability that an animal on the Hochstatter farm is a brown goat?

   What is A + B + C? (2006 MathIsCool)

2. Three members of the Euclid Middle School girls' softball team had the following conversation.
   Ashley: I just realized that our uniform numbers are all 2-digit primes.
   Brittany : And the sum of your two uniform numbers is the date of my birthday earlier this month.
   Caitlin: That's funny. The sum of your two uniform numbers is the date of my birthday later this month.
   Ashley: And the sum of your two uniform numbers is today's date.
   What number does Caitlin wear? (2014 AMC8)

3. Five friends compete in a dart-throwing contest. Each one has two darts to throw at the same circular target, and each individual's score is the sum of the scores in the target regions that are hit. The scores for the target regions are the whole numbers 1 through 10. Each throw hits the target in a region with a different value. The scores are: Alice 16 points,

Ben 4 points, Cindy 7 points, Dave 11 points, and Ellen 17 points. Who hits the region worth 6 points? (2004 AMC8)

4. The numbers 2, 4, 6, 9, and 12 are rearranged according to these rules:
   1. The largest isn't first, but it is in one of the first three places.
   2. The smallest isn't last, but it is in one of the last three places.
   3. The median isn't first or last.
   What is the average of the first and last numbers? (2004 AMC8)

5. The six children listed below are from two families of three siblings each. Each child has blue or brown eyes and black or blond hair. Children from the same family have at least one of these characteristics in common. Which two children are Jim's siblings? (2003 AMC8)

Child	Eye Color	Hair Color
Benjamin	Blue	Black
Jim	Brown	Blonde
Nadeen	Brown	Black
Austin	Blue	Blonde
Tevyn	Blue	Black
Sue	Blue	Blonde

6. Kaleana shows her test score to Quay, Marty and Shana, but the others keep theirs hidden. Quay thinks, "At least two of us have the same score." Marty thinks, "I didn't get the lowest score." Shana thinks, "I didn't get the highest score." List the scores from lowest to highest for Marty (M), Quay (Q) and Shana (S). (2001 AMC8)

7. Five runners, P, Q, R, S, T, have a race, and P beats Q, P beats R, Q beats S, and T finishes after P and before Q. Who could NOT have finished third in the race? (1993 AMC8)

8. Abby, Bret, Carl, and Dana are seated in a row of four seats numbered #1 to #4. Joe looks at them and says:
   "Bret is next to Carl."
   "Abby is between Bret and Carl."
   However each one of Joe's statements is false. Bret is actually sitting in seat #3. Who is sitting in seat #2? (1987 AMC8)

9. Alan, Beth, Carlos, and Diana were discussing their possible grades in mathematics class this grading period. Alan said, "If I get an A, then Beth will get an A." Beth said, "If I get an A, then Carlos will get an A." Carlos said, "If I get an A, then Diana will get an A." All of these statements were true, but only two of the students received an A. Which two received A's? (1986 AMC8)

Number Base

Many different number bases are being used in our daily lives. Their existences have good reasons. We need to understand and master how to convert numbers among different bases. In this section, we will focus on learning how to convert numbers between decimal (base 10) and binary (base 2).

Decimal to Binary

[Example]
Convert number 350 in base 10 to a binary number (base 2)

[Solution]
In base 10, we can write 350 in this equation:
$$350 = 3 * 10^2 + 5 * 10^1 + 0 * 10^0$$

Notice each coefficient (i.e. 3, 5, 0) is less than 10, and there is no coefficient for 10^3 or above.

Now we want to change it to something like this:
$$350 = a * 2^8 + b * 2^7 + c * 2^6 + d * 2^5 + e * 2^4 + f * 2^3 + g * 2^2 + h * 2^1 + i * 2^0$$

Notice there is no 2^9 or above, because we know $350 < 512 = 2^9$

$350 - 1 * 256$ (i.e. 2^8) $= 94 < 128 = 2^7$ → a = 1, b = 0;
$94 - 1 * 64$ (i.e. 2^6) $= 30 < 32 = 2^5$ → c = 1, d = 0;
$30 - 1 * 16$ (i.e. 2^4) $= 14$ → e = 1;
$14 - 1 * 8$ (i.e. 2^3) $= 6$ → f = 1;
$6 - 1 * 4$ (i.e. 2^2) $= 2$ → g = 1;
$2 - 1 * 2$ (i.e. 2^1) $= 0$ → h = 1, i = 0;

Therefore, $350 = 1 * 2^8 + 0 * 2^7 + 1 * 2^6 + 0 * 2^5 + 1 * 2^4 + 1 * 2^3 + 1 * 2^2 + 1 * 2^1 + 0 * 2^0$

Which means, $(350)_{10} = (101011110)_2$

The subscript number "10" and "2" indicates its number base.

Binary to Decimal

[Example]
Convert binary number 11001001 to a decimal number

[Solution]

We rewrite the expression of the binary number as below.

$(11001001)_2 = 1 * 2^7 + 1 * 2^6 + 0 * 2^5 + 0 * 2^4 + 1 * 2^3 + 0 * 2^2 + 0 * 2^1 + 1 * 2^0$

$= 128 + 64 + 8 + 1 = (201)_{10}$

Fraction in Decimal to Binary

We need to learn how we can identify each digit after the decimal point. For example, a fraction number 4.3256, in ten base.

Remove integer part "4", we have 0.3256.
   0.3256 x 10 = 3.256 → 3 is the 1st digit after decimal point
Remove integer part "3", we now have 0.256
   0.256 x 10 = 2.56 → 2 is the 2nd digit after decimal point
Remove integer part "2", we now have 0.56
   0.56 x 10 = 5.6 → 5 is the 3rd digit after decimal point
Remove integer part "5", we now have 0.6
   0.6 x 10 = 6 → 6 is the 4th digit after decimal point
Remove integer part "6", we are done.
The same process applies when we convert a fraction from decimal to binary.

Integer part of "4.3256" is "4", which is 100 in binary.
From now on, we only look at decimal part.
   0.3256 x 2 = 0.6512 → 0 is the 1st digit after decimal point
   0.6512 x 2 = 1.3024 → 1 is the 2nd digit
   0.3024 x 2 = 0.6048 → 0 is the 3rd digit
   0.6048 x 2 = 1.2096 → 1 is the 4th digit
   0.2096 x 2 = 0.4192 → 0 is the 5th digit
   0.4192 x 2 = 0.8392 → 0 is the 6th digit
   0.8392 x 2 = 1.6784 → 1 is the 7th digit

   ......
Repeat until we finally get 0, or any repeating pattern.

$(100.0101001...)_2$ would be the equivalent fraction number in base two, or binary.

Binary Arithmetic

It includes Addition, Subtraction, Multiplication, Division, Square root

   Binary addition and subtraction operation follows rules as:
   0 + 0 = 0 → 0 - 0 = 0
   0 + 1 = 1 → 1 - 0 = 1

1 + 0 = 1
1 + 1 = 10 → 10 - 1 = 1

Binary multiplication and division operation follows rules as:
0 x 0 = 0
0 x 1 = 0
1 x 0 = 0
1 x 1 = 1

This is an example of division operation between two binary numbers.

```
       11
11) 1011
    -11
    101
    -11
     10      →      remainder (r)
```

[Example]
In what base is 204 + 314 = 521? (2000 MIC 6G/37)

[Solution]
The number base must be greater than any digit of the numbers. It should be greater than 5.
The sum operation of the unit digits is 4 + 4 = 1, has implied the base is 7.
Validated the equation with base 7 → confirmed the answer is 7.

[Example]
What is the product (in base 5) of the following: $2 \times 123_5$? (2005 MathIsCool)

[Solution]
$2 \times (1 \times 5 \times 5 + 2 \times 5 + 3) = 2 \times 5 \times 5 + 4 \times 5 + 5 + 1 = 2 \times 5 \times 5 + 5 \times 5 + 1 = 3 \times 5 \times 5 + 0 \times 5 + 1 = 301_5$

[Practice Problems]

1. Write the number 2007 in the smallest base for which the unit's digit would remain 7. Include the base in your answer as a subscript. (2007 MathIsCool)
2. When the base 10 number 2013 is written in base 2, how many digits will it have? (2013 MathIsCool)
3. What is the probability a randomly generated four digit base 2 number will be greater than 12_8? (2007 MathIsCool)

Algebraic Way of Thinking

So far, we have mostly used arithmetic approaches to solve many different types of word problems. As we explore more mathematical ideas with more challenging problems, we will find Algebra is a much more powerful tool to solve elementary math problems, once they are well-modeled.

What is Algebra?

Algebra is a mathematical modeling expression that uses letters along with numbers. The letters, a.k.a. variables, stand for numbers that are unknown.

Play with Algebra

Simplifying Expressions
- An expression is a mathematical statement that may use numbers, variables, or both.

[Example]
Simplify: $2(43x + 2) - 86x + 2$

[Solution]
Combine liked items ➜ $86x + 4 - 86x + 2 = 6$.

Evaluating Expressions
- Substitute the value of the variable into the expression and computing the result.

[Example]
Evaluate: $39x + 12(x+1) + 75(x-1)$, when $x=2$

[Solution]
Substitute x with 2 ➜ $39 \times 2 + 12(2+1) + 75(2-1)$ ➜ $78 + 36 + 75 = 189$.

Solving for a Variable
- Find value for a variable from an equation.

[Example]
Find x, if $9x + 7 = 34$

[Solution]
Add -7 to both sides of the equation ➜ $9x + 7 - 7 = 34 - 7$

Simplify them ➜ 9x = 27
Divided by 9 ➜ 9x / 9 = 27 / 9
Solve x = 3

[Example]
x+y+z = 10. If x=3 and y=4, what does z equal?

[Solution]
Replace x, y with 3, 4 respectively ➜ z + 7 = 10
Subtract 7 from both sides ➜ z + 7 − 7 = 10 − 7
Answer is: z = 3

The beauty in algebraic methods is that, regardless of known data or unknown variables, as long as you can build math equations to reveal the underneath relationship among them, you are almost guaranteed to find solutions. There are two key steps here.
1. Identify all the key data points.
    Some of them may already be clear from the given data. Some of them are unknown, but you can define variables, like x, y, z to represent them temporarily.
2. Establish math equations using the key data points.
    This is called "math-modeling", which is very important and powerful methodology.
3. Solve equations and find out values for the variables.

[Example]
Ron and Hermione are on a road trip. They decide to split the driving: for every three miles Ron drives, Hermione drives five miles. Ron averages 30 miles per hour (mph) and Hermione averages 80 mph. What is the overall average speed for the trip, in mph? (2011 MathIsCool Master 4G/40)

[Solution]
Step 1. Define total travel has d miles, and the overall average speed is x mph.
        Then, Ron drove 3d/8, Hermione drove 5d/8.
Step 2. According to Distance = Speed x Time, we build the following equation,
$$\left(\frac{3d/8}{30} + \frac{5d/8}{80} \right) x = d$$
Step 3. After reducing factors from both side, we get, $\left(\frac{1}{80} + \frac{1}{128} \right) = x$

Conclusion ➜ The overall average speed is, x = 640 / 13 = 49 mph.

[Example]
Biff ran 4 times as fast as Eho. In fact, he ran 80 miles in 2 hours less than it took Eho to run 28 miles. How fast did Eho run? (2007 MathIsCool)

[Solution]
Step 1. Define Eho's speed x mph, so Biff's speed is 4x mph
Step 2. Establish equation ➜ $80/(4x) + 2 = 28 / x$
Step 3. Solve equation ➜ $20/x + 2 = 28/x$
　　　　Move 20/x to right side ➜ $2 = 28/x - 20/x$ ➜ $2 = 8/x$ ➜ $x = 4$
Conclusion ➜ Eho can run 4 miles per hour.

[Example]
Loki, Moe, Nick and Ott are good friends. Ott had no money, but the others did. Moe gave Ott one-fifth of his money, Loki gave Ott one-fourth of his money and Nick gave Ott one-third of his money. Each gave Ott the same amount of money. What fractional part of the group's money does Ott now have? (2002 AMC8)

[Solution]
Step 1. Assume before they gave money to Ott,
　　　　Loki has L dollar, Moe has M dollar, Nick has N dollar.
Step 2. Build equation ➜ $M/5 = L/4 = N/3 = t$, a new variable is defined for convenience later.
Step 3. Transform the equation
　　　➜ The group has $M + L + N = 5t + 4t + 3t = 12t$, amount of money (dollars).
　　　　Now, Ott has got $M/5 + L/4 + N/3 = 3t$, amount of money (dollars)
Conclusion ➜ Ott now has $3t / (12t) = ¼$ of the group's money.

[Example]
A goat was born in the year x^2 and died on her 84th birthday in the year $(x+2)^2$. What year was the goat born? (2006 MathIsCool)

[Solution]
Step 1. Variables are defined in terms of x.
Step 2. Establish equation ➜ $(x+2)^2 - x^2 = 84$
Step 3. Solve equation ➜ expand left side and simply it: $4x + 4 = 84$ ➜ $4x = 80$ ➜ $x = 20$
Conclusion ➜ the goat was born in year $20^2 = 400$.

[Example]
The average weight of my dog, my cat, and my rabbit is 29 pounds. The average weight of my dog and my cat is 41 pounds. How many pounds does my rabbit weigh? (2007 MathIsCool)

[Solution]
Step 1. Define variables ➜ the weight of dog, cat and rabbit are D, C, and R, respectively.
Step 2. Build equations ➜ $D + C + R = 3 \times 29$, and $D + C = 2 \times 41$
Step 3. Find out R ➜ $3 \times 29 - 2 \times 41 = 87 - 82 = 5$

89

Conclusion ➜ My rabbit weighs 5 pounds.

[Practice Problems]

1. Ralph went to the store and bought 12 pairs of socks for a total of $24. Some of the socks he bought cost $1 a pair, some of the socks he bought cost $3 a pair, and some of the socks he bought cost $4 a pair. If he bought at least one pair of each type, how many pairs of $1 socks did Ralph buy? (2015 AMC8)

2. Jeremy's father drives him to school in rush hour traffic in 20 minutes. One day there is no traffic, so his father can drive him 18 miles per hour faster and gets him to school in 12 minutes. How far in miles is it to school? (2015 AMC8)

3. The manager of a company planned to distribute a $50 bonus to each employee from the company fund, but the fund contained $5 less than what was needed. Instead, the manager gave each employee a $45 bonus and kept the remaining $95 in the company fund. What was the amount of money in the company fund before any bonuses were paid? (1996 AMC8)

4. In a river with a steady current, it takes a frog 20 minutes to swim a certain distance upstream, but it takes her only 10 minutes to swim back. How many minutes would it take a stick to float this same distance downstream? (2006 MathIsCool)

5. If P pizzas can be purchased for D dollars, how many cents would be necessary to purchase 3 pizzas? (2007 MathIsCool)

6. Silas has a collection of foreign coins worth 8 cents and 5 cents. What is the largest price of something he could not pay for exactly, using 8-cent and 5-cent coins? (2001 MathIsCool)

7. Seventy-five marbles are placed in boxes 1,2,3,4, and 5. Boxes 1 and 2 contain a total of 27 marbles. Boxes 2 and 3 contain a total of 25 marbles. Boxes 3 and 4 contain a total of 31 marbles. Boxes 4 and 5 contain a total of 38 marbles. How many marbles are in box 1? (2007 MathIsCool)

8. Cam and Joel each had at least one apple. Cam gave half of his apples to Joel. Then Joel gave two-thirds of all his apples to Cam. Cam now has 10 apples. How many apples do they have altogether if only whole apples are exchanged? Give all possible answers. (2008 MathIsCool)

9. At a classroom costume party, the average age of the b boys is g, and the average age of the g girls is b. If the average age of everyone at the party (all these boys and girls, plus their 42-year-old teacher) is b+g, what is the value of b+g? (2004 MathIsCool)

10. On a plane, two men together had 135 kilograms of luggage. The first paid $1.35 for his excess luggage and the second paid $2.70 for his excess luggage. Had all the luggage belonged to one person, the excess luggage charge would have been $8.10. At most how many kilograms of luggage is each person permitted to bring on the plane free of additional charge? (2004 MathIsCool)

11. Two standard 6-sided dice are rolled. The probability of rolling a sum of x is 1/12. Find the product of all possible integers of x. (2004 MathIsCool)

12. Abe pushed a boulder up a 40 foot hill. He pushes it 12 feet each day, and each night it rolls down x feet, where x is an integral value. Abe reached the top of the hill with the boulder on the 7th day. What is the sum of all the possible values of x? (2002 MathIsCool)

13. Circles A and B are concentric, and a chord of circle B that is tangent to circle A measures 14 m. What is the area, in square meters, of the region between the two circles? (2007 MathIsCool)

14. In a jar of red, green, and blue marbles, all but 6 are red marbles, all but 8 are green, and all but 4 are blue. How many marbles are in the jar? (2012 AMC8)

15. Ruthie has 10 coins, all either nickels, dimes, or quarters. She has N nickels, D dimes, and Q quarters, where N, D, and Q are all different, and are each at least 1. Amazingly, she would have the same amount of money if she had Q nickels, N dimes, and D quarters. How many cents does Ruthie have? (2012 MathIsCool)

16. Some workers were asked to mow two fields, one twice as big as the other. They all mowed the larger field for half of a day; then they split in half. One group finished the larger field at the end of the day. The others mowed the smaller field; but at the end of the day, there remained a part to do. This part was finished by one worker in a single day. How many workers were there? (2004 MathIsCool)

17. When a certain counting number is divided by 9, the sum of the remainder and the quotient is 13. What is the smallest possible value of this number? (2010 MathIsCool)

18. Starting with some gold coins and some empty treasure chests, I tried to put 9 gold coins in each treasure chest, but that left 2 treasure chests empty. So instead I put 6 gold coins in each treasure chest, but then I had 3 gold coins left over. How many gold coins did I have? (2017 AMC 8)

Newton's Cows and Grass

[Example]
The daily vertical growth rate and the per square area density of the grass on a grazing field are constant. 70 cows consume the entire grass on the entire field in 24 days, while 30 cows do it in 60 days. How many cows does it take to consume the entire grass on the entire field in 96 days?

[Solution]
The key is to find out the volume of grass grow daily. Then you get to know the original grass volume.
Grass grows every day by: $(30 \times 60 - 70 \times 24) / (60 - 24) = (1800 - 1680) / 36$
$= 120 / 36 = 10 / 3$ (cow's grass per day).
Initial grass volume: $24 \times 70 - 24 \times 10/3 = 1680 - 80 = 1600$ (cow's grass).
Assuming it will take x cows,
$1600 + 96 (10/3) = 96x$ ➔ $x = (1600 + 320) / 96 = 1920 / 96 = 20$ days.

[Practice Problems]

1. In a field, 17 cows can finish the whole grass in the field for 30 days. 19 cows can finish in 24 days. If a group of cows eat the grass for 6 days, then 4 cows are sold, the remaining cows can finish the grass in 2 days. What is the initial number of cows before 4 of them are sold?

Numbers and Words

Numbers don't only belong to Math. Math is not just with numbers. Numbers exist everywhere among the words in any languages. Math is a special language that consists of numbers and other symbols. There is at least one interesting story behind every single commonly-used number. It will be very helpful for math learners to know more about what numbers mean in the real world.

In this chapter, you will find out what some of the most frequently used numbers really mean to us in our lives. Please try the following puzzles by completing the words in accordance with what the number implies. Among the equalities below, you only get the first letter as a hint for each word of which you are supposed to figure out the full spelling.

For instance,

"5 = F on each H" means "5 Fingers on each Hand";

"60 = M in an H" indicates "60 Minutes in an Hour".

Then, what about?

- 3 => S on a T
- 4 => S in a Y
- 5 => T on each F
- 6 => S on a C
- 7 => D in a W
- 8 => S on a S S
- 9 => P in the S S
- 12 => E on a C
- 13 => C in a S
- 18 => H in a G C

- 24 => H in a D

- 26 => L of the A

- 29 => D in F in a L Y

- 52 => C in a D (without the J)

- 54 => C in a D (with the J)

- 60 => S in a M

- 64 => S on a C B

- 88 => P K

- 90 => D in a R A

- 206 => B in the B

- 360 => D in a C

- 747 => B A

- 1001 => A N

- 1040 => T R F

- ……

I hope you have fun and learn something more about these numbers by this practice. After you have worked out some of them, you may also want to share with you friends or others. You can find all of the answers on the last page. Would you like to add more examples of this kind of puzzle involving both words and numbers?

Answer Keys

Clock

1. 3:52 + 12:00 = 8:10 = 7 hr 42 min

2. 1000 − 12 x 83 = 4. 2 + 4 = 6 o'clock.

3. 1 + 2 + … + 12 = 78.

4. LCM = (3, 4, 6) = 12 seconds after 12 o'clock.

5. 360 * (5 / 12) = 150 degree

6. The smaller angle between number 4 and 5 on the clock is 30 degrees. By 4:20, the hour hand would have moved 20/60 way from 4 to 5, which is one-third of 30 degrees. Since the minute hand is pointing to 4 at the moment, therefore the answer is 30 / 3 = 10 degrees between the two hands.

7. 70 degrees

8. 12 * (60 / 64) = 45/4 hours, which is 11:15 PM

9. It is 264 minutes from 3 PM to 7:24 PM. 264 * (60 / 40) = 396 minutes, or 6 hr 36 min. Therefore, it is 9:36 PM.

10. Alice moves 5 * n steps and Bob moves 9 * n steps, where n is the # of turn they are on. They will meet when 5 * n + 9 * n = 14 * n, is a multiple of 12. The smallest number of turns is 6.

11. Hint: We know the hour hand moves one minute's space whenever the minute hand moves 12 minutes' space.

12. When it is at 11th, or 12th, or 21st, 22nd minute of 1, or 2, or 11, or 12 o'clock, each display lasts one minute. There are 4 x 4 = 16 minutes, in total.

13. From 1 to 9 o'clock, the times will be of the form a:ba , There are 9 choices for a (1 to 9) and 6 choices for b (0 to 5), so there are 9 * 6 = 54 times in this period. From 10 to 12 o'clock, the minutes are fixed, i.e. 10:01, 11:11, 12:21, so there are only 3 times in this case. In total, there are 54 + 3 = 57 times.

Money

1. smallest possible total number of coins in all the purses,
    Purse A: 1Q, 0D, 0N, 3P → 4
    Purse B: 0Q, 2D, 1N, 3P → 6
    Purse C: 0Q, 1D, 3N, 3P → 7
    Purse D: 0Q, 0D, 5N, 3P → 8
    Purse E: 0Q, 2D, 0N, 8P → 10
    (Q: quarter, D: dime; N: nickel, P: penny)
    4 + 6 + 7 + 8 + 10 = 35 coins in total.

2. (100 − 16) / 14 = 6 months.

3. 2, or 6, or 10, or 14, or 18 dimes.

4. (6 x 24 − 119) / (6 − 3.5) = 10.
    10 X 3.5 x 40% = $14

5. 1200 − 452 = 748 = 82 x (15 − 6) + 5 + 5, which means 3 x 82 + 1 + 1 = 248 days.

6. Least common multiple problem. The LCM of 50, 70, and 75 is 1,050 dimes, which is $105.

7. (#Dimes, #Nickels, #Pennies) =
    (5, 0, 2) − 1 case,
    (4, 2, 2), (4, 1, 7), (4, 0, 12) − 3 cases,
    (3, 4, 2), (3, 3, 7), (3, 2, 12), (3, 1, 17), (3, 0, 22) − 5 cases,
    (2, 6, 2), … − 7 cases,
    (1, 8, 2), … - 9 cases,
    (0, 10, 2), …− 11 cases,
    1 + 3 + 5 + 7 + 9 + 11 = 36 different ways in total.

8. 2 x 50 − (0.6 x 50 + ½ x 50) = 45. He saved 45/150 = 30%.

9. The cost must be a factor of both 143 and 195. They have common factor 1 and 13. It
    cannot be 1, or there would have to be more than 30 sixth graders. The pencil has to cost 13
    cents. There are (195 − 143) / 13 = 4 more sixth graders.

10. After the 15% loss, Tammy has 100 x 0.85 = 85 dollars. After the 20% gain, she has 85 x 1.2 =
    102 dollars. The change has (102 − 100) / 100 = 2% gain.

11. (500 − 50 − 35 − 15) / 16 = 25 weeks.

12. $40 x (1 + 10%) x (1 – 10%) = $39.60.

13. 1 – (1 – 30%) x (1 – 20%) = 1 – 56% = 44%.

14. The nine coins are three quarters, one dime, three nickels, and two pennies.

15. Toy's money change: $36 → $72 → $144 → $36, at his last distribution, we see total amount is actually 36 + (144 – 36) x 2 = $252.

16. ($5/4 – $4/5) / ($5/4) = 0.45 / 1.25 = 36% decrease.

17. $2,000 x 1.2 x 0.8 = $1,920.

18. ($100 x (1 – 20%) – $5) x (1 + 8%) = $81.

19. 128 x (1 – 25%) x (1 – 25%) – 1 = 71 apples.

20. $180 x (1 – 50%) x (1 – 20%) = $72.

21. Ten coins: 4 pennies, 1 nickel, 2 dimes, and 3 quarters.

Travel

1. During those 10 seconds, the sound traveled feet from the lightning to Snoopy. This is equivalent to 1088 * 10 / 5280 = 2 miles, roughly.

2. Answer key is Evelyn, according to:
Average speed = distance / time.

3. (95 - 14) / 3 = 27 minutes, 4 * 27 - 78 = 30 units.

4. 5 + 8 = 13 is 1/3 of the total distance, so it is 13 / (1/3) = 39 miles from Biff's house to Eho's house.

5. Alex bikes at a rate of 8 / 1.25 = 32/5 meter/second. Randy is 500 meters behind Alex. For Randy to catch up with Alex, it will take 500 / (8 - 32/5) = 2500 / 8 seconds. (2500 / 8) * 8 / 800 = 3.125 laps.

6. 4 – 5/2 = 3/2 ft.
40 = (3/2) x 24 + 4 → 25 days for the snail to get out of the well.

7. Each lap Bonnie runs, Annie runs one quarter lap more. Bonnie will run four laps before she is first time overtaken by Annie. At that time, Annie will have run 5 laps.

8. Diameter of the semi-circle = 40 ft.
   Total number of semi-circles = 5280 / 40 = 132
   Circumference of one semi-circle = 40 x Pi / 2 = 20 x Pi ft.
   132 x 20 x Pi (ft) / 5 (mph) = Pi (mile) / 10 (mph) = Pi/10 (hour).

9. On the 11th jump, the Hulk jumps 1024 meters, first time exceeding 1000 meters (1 km), so the answer is the 11th jump.

10. The actual time he spent on treadmill is: 2/5 + 2/3 + 2/4. If he always walked at 4 mph, the total time would be 2/4 + 2/4 + 2/4. Their difference is 1/15 hours. He would have spent 4 minutes less time on treadmill.

11. (½ + ½) / (12 – 8) = ¼ hours = 15 minutes.

12. 50 * 2 / (50 / 60 + 50 / 40) = 48 mph.

13. (62 – 12 x (1/2)) / (12 + 16) = 2 hrs. It is 11:00 a.m. when they meet.

14. 85 x 9 = 765 minutes. 765 – 75 x 5 – 90 x 3 = 120 minutes = 2 hours.

15. There are 160 – 40 = 120 miles between the third and tenth exits. The service center is at milepost (40 + 120 * ¾) = 130.

16. 5 / (1/2) = 10. All the passes occur in the middle of highway.

17. (80 + 100)/4 = 45 mph.

18. (57,060 – 56,200) / (12 + 20) = 26.9 mph, roughly.
    Read carefully and identify what information is irrelevant to the solution.

19. It will decrease in height every time it bounces back. At 1st time: 2 m; 2nd time: 4/3 meter; 3rd time: 8/9 m; 4th time: 16/27 m; 5th time: 32/81 < ½ m. Therefore, On 5th bounce, it will rise to a height less than 0.5 meters.

20. (100 -2) Pi + (60 + 2) Pi + (80 – 2) Pi = 238 Pi.

21. 25 miles.

Four-legs vs. Two-legs
1. There were three apples and two oranges, since the total weight was a whole number of pounds. 3 x ½ x 0.8 + 2 x ¾ x 1.1 = $2.85.

2. $(45 \times 6 - 230) / (6 - 4) = 20$ $4 tickets.

3. $(87 \times 10 - 602) / (10 - 6) = 67$ six-point bucks.

4. $(30 - 2 \times 12) / (4 - 2) = 3$ elephants.

5. $(62 - 19 \times 2) / (4 - 2) = 12$ camels.

6. $(2200 \times 4 - 5050) / (4 - 1.5) = 1500$ children.

7. One set of table and chairs (including one table and four chairs) has 17/5 legs.
   $(18 \times 17/5 - 60) / (17/5 - 3) = 3$ stools.

8. $(200 \times 4 - 522) / (4 - 2) = 139$.

9. $(19 - 2 \times 7) / (3 - 2) = 5$.

10. $(29 - (-2) \times 10) / (5 - (-2)) = 7$.

11. The highest possible score is 100. The next highest is 96. (E) 97 is impossible.

12. Maximum number of questions is 12.

Age
1. Drawing two lines, one for Hanson's age after 16 years, and the other is for his age ten years ago. You will find Hanson's current age is $(10 + 16) + 10 = 36$.

2. Henry's current age is $20 / (5 - 1) = 5$ years old.

3. 54

Work
1. $4 \times 15 / 10 = 6$ days.

2. After the 1st 4 minutes, Homer had peeled $4 \times 3 = 12$ potatoes. Once Christen joined him, the two are peeling potatoes at a rate of $3 + 5 = 8$ potatoes per minute. So, they finished peeling after another $(44 - 12) / 8 = 4$ minutes. In these 4 minutes, Christen peeled $4 \times 5 = 20$ potatoes.

3. $1 / (1/50 - 1/90) = 50 \times 90 / 40 = 112.5$ seconds.

4. $(1/2) / (1/3 - \frac{1}{2} / 2) = 6$ mph.

5. Battery used after nine hours is $1/3 + 8 \times (1/24) = 2/3$.
   $(1 - 2/3) / (1/24) = 8$ hours left to be left on.

6. [A]

7. Define the distance between Joe's home and his school is 1.
   Joe's running speed: $(1/2) \times 3 / 6 = \frac{1}{4}$.
   Joe's time spent on running: $(1/2) / (1/4) = 2$ minutes.
   $6 + 2 = 8$ minutes.

Mixture

1. The first pitcher contains $600 \times 1/3 = 200$ mL of orange juice. The second pitcher contains $600 \times 2/5 = 240$ mL of orange juice. In the large pitcher, there is a total of $200 + 240 = 440$ mL of orange juice and $600 + 600 = 1200$ mL of fluids, giving a fraction of $440/1200 = 11/30$.

2. A pear gives $8/3$ ounces of juice per pear. An orange gives $8/2 = 4$ ounces of juice per orange. If the pear-orange juice blend used one pear and one orange each, the percentage of pear juice would be $8/3$ $(8/3 + 4) = 8 / (8 + 12) = 40\%$.

3. Sugar total: $10 + 40 \times 30\% = 22$ g. Mix total: $10 + 40 = 50$ g. $22/50 = 44\%$ is the answer.

Painted Cubes

1. The surface area of the cube is $6 \times 3 \times 3 = 54$. Each of the eight black cubes has 3 faces on the outside, making $3 \times 8 = 24$ black faces. There are $54 - 24 = 30$ white faces, which is $30/54 = 5/9$ of the whole surface area.

2. The original cube has 6 faces, with surface area of $2 \times 2 = 4$ square units. Thus, the original figure had a total surface area of 24 square units. The new figure has the original surface, with 6 new faces that each has an area of 1 square unit, for a total surface area of 6 additional square units added to it. But 1 square unit of the top of the bigger cube, and 1 square unit on the bottom of smaller cube, is not on the surface. They do not count towards the surface area.

   The total surface area is therefore $24 + 6 - 1 - 1 = 28$ square units.

   The percent increase in surface area is $(28 - 24) / 24 = 16.7\%$

3. $4 + 5 + 4 + 5 + 2 + 2 = 22$ faces.

4. Assuming Marla evenly distributes her 300 square feet of paint between the 6 faces, each face will get 300/6 = 50 square feet of green paint. The surface area of one cube face is 10 x 10 = 100 square feet. There will be 100 – 50 = 50 square feet of white on each side

5. Think about the opposite question, "How many cubes are not touching a side or the bottom of the box?" Using supplementing approach, 64 – (4 + 4 + 4) = 52 small cubes.

6. If a green square cannot share its top or right side with a red square, then a red square cannot share its bottom or left side with a green square. Let us do case work.

   When there are no green squares ➜ one way;

   When there is one green square and three red squares ➜ one way;

   When there are two green squares and two red squares ➜ two ways;

   When there are three green squares and one red square ➜ one way;

   When there are four green squares and zero red squares ➜ one way;

   In total, 1 + 1 + 2 + 1 + 1 = 6 ways.

7. The sides of the pyramid contribute 4 + 8 + 12 = 24 for the surface area. The total top surface area is 9 – 4 + 4 – 1 + 1 = 9. In total, it is 24 + 9 = 33 square meters.

Spatial Thinking

1. 0 cm.

2. 9, 10, and 15.

3. We undo step 4 back to step 1, and discover that 9 is the right answer.

4. Originally there are 12 edges. Newly added edges are 3 x 8 = 24. There will be 36 edges after the cut.

5. Faces 6, 5, and 4 cannot come together as a corner. We find the faces 6, 5, and 3 are together at a common vertex. The largest sum will be 14.

6. Referring to the diagram, the perimeter of one of the smaller rectangles is 10x, and the perimeter of the large rectangle is 12x. The ratio is (10x) / (12x) = 5/6.

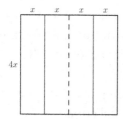

7. Answer is (A), if you keep track of the "bottom" side of the square.

Pattern Recognition

1. 7 x 7 – 6 x 6 = 13 more tiles.

2. We may start by looking for a pattern, 98, 49, 44, 22, 11, 6, 54, 27, 22, 11, 6, … We see a pattern of 22, 11, 6, 54, 27, … after 98, 49, 44. There are 5 terms in each repetition of the pattern, and 95 = 0 (mod 5), so the answer is 27.

3. All small triangles are congruent in each iteration of the diagram. Notice the first eight terms are triangular numbers, 0, 1, 3, 6, 10, 15, 21, 28. In the eighth diagram, there will be 28 shaded triangles. The eighth triangle will be divided into 64 small triangles. The ratio of shaded triangles to total triangles will be 28/64 = 7/16.

4. The numbers shaded are the triangular numbers, which are numbers in the form $n(n + 1)/2$ for positive integers. Squares that have the same remainder after being divided by 8 will be in the same column. Thus, we want to find when the last remainder, from 0 to 7, is found. Note that the triangular numbers up to 120 are 1, 3, 6, 10, 15, 21, 28, 36, 45, 55, 66, 78, 91, 105, 120. When you divide each of those numbers by 8, all remainders must be present. The last one to fill in all columns is 120.

5. Regroup numbers together with their operations as below:
    (1901 – 101) + (1902 – 102) + (1903 – 103) + … + (1993 – 193) = 1800 x 93 = 167,400.

6. The last number in each row ends in a perfect square. Thus 142 is two left from the last number in its row, 144. One left and one up from 144 is the last number of its row is 121. The number directly above 142 is 120.
    … 120 121
    … 142 143 144

7. The sequence consists of repeated strings of 9 letters, ABCDEDCBA. When 1992 is divided by 9, the remainder is 3. So the 1992nd letter in the sequence is the 3rd letter of the smaller string, C.

8. (D)

9. From the math expression, total number of factor 5 is 14 and total number of factor 2 is 9. A pair of factor 2 and factor 5 contributes to one zero. There will be 9 zeros in the final product result.

10. With each change, ¾ of the black space from the previous stage remains. Since there are 5changes, the fractional part of the triangle that remains black is ¾ to the 5th power. The answer is 243/1024.

11. First number is ¼

12. There are eight distinct squares in total.

13. Every 4th line has 1989 as part of it and every 6th line has AJHSME as part of it. In order for both to be part of line n, n must be a multiple of 4 and 6, When n = LCM(4, 6) = 12, it will appear for the first time.

14. 210.

Three Cats

1. 21 Haps = 7 Hups ➔ 42 haps = 14 Hups;
   14 Hups = 4 Hops ➔ 42 Haps = 4 Hops ➔ 21 Haps = 2 Hops ➔ 84 Haps = 8 Hops;
   28 Hips = 16 Haps ➔ 7 Hips = 4 Haps ➔ 147 Hips = 84 Haps ➔ 147 Hips = 8 Hops;
   1   Hip = 8/7 dollar ➔ 1 Hop = 147/8 Hips = (147/8) x (8/7) = $21.

2. 24T = 18Z ➔ 20T = 15Z = 3F ➔ 40T = 6F= 8B ➔ 1B = 5T, 5 Twinks in a Blink.

3. 90 wazzas.

4. 11. There are 6 boinks in one bunch and 2 blinks in 10 bunches.

5. 2 x 1500 x (1 + 1 x 8 + 1 x 8 x 17) = 435,000 moons, planets, and starts.

6. Among 5 x 4 = 20 durians, trade 4 durians for 12 apricots, trade 3 durians for 6 peaches, and trade 3 durians for 6 oranges. Ton can make 6 smoothies, with one durian remaining.

7. The number of 8th graders has to be a multiple of 8 and 5. The smallest possible number of 8th graders is 40. Then there are 40 x 3/5 = 24 6th graders and 40 x 5/8 = 25 7th graders. The numbers of students is 40 + 24 + 25 = 89.

8. Only considering the chocolate Jordan has, it can make up to 5 x 5/2 = 12.5 servings, the sugar can make up to 2 x 5 / (1/4) = 40 servings, the water can make unlimited servings, and

the milk can make up to 7 x 5 / 4 = 8.75 servings. Limited by the amount of milk, Jordan can make at most 8.75 servings.

9. 3 fish ➔ 2 loaves of bread; one loaf of bread ➔ 4 bag of rice; 3 fishes ➔ 8 bags of rice. So each fish is worth 8/3 bags of rice.

Count

1. All single M&M: 4! / 2! = 12;
   One double M&M: 3 x 2 x 3 + 3 = 21;
   Two double M&M: 2 + 2 = 4;
   In total, 12 + 21 + 4 = 37 orders.

2. 4! / 2! = 12.

3. 16 shoes.

4. If a person gets three gumballs of each of the three colors, then the 10th gumball must be the fourth one for one of the colors. Therefore, the person must buy 10 gumballs.

5. Because 911 is not allowed, for the first three digits, there are 10^3 -1 = 999 combinations for the first three digits. In total, it will have 999 x 10 = 9990 possible passwords.

6. Combination for (3-1) out of (3 + 5 − 1), which is 21.

7. C(4+4−1, 4−1) = C(7, 3) = 35.

8. C(5+5-1, 5-1) = C(9, 4) = 126.

9. Total number of outfits: 3 x 5 x 4 x 14 = 840;
   Total with yellow hat and yellow shirt (but no yellow skirt): 1 x 1 x (4 − 1) x 14 = 42;
   Total with yellow hat and yellow skirt (but no yellow shirt): 1 x (5 -1) x 1 x 14 = 56;
   Total with yellow skirt and yellow skirt (but no yellow hat): (3 − 1) x 1 x 1 x 14 = 28;
   Total with yellow hat, yellow shirt, and yellow skirt: 1 x 1 x 1 x 14 = 14.
   Doris can make 840 − 42 − 56 − 28 − 14 = 700 different outfits.

10. 78 − 14 − 1 = 63. 78 + 63 x 2 + 1 = 205th number person in line.

11. With one "0" in the middle: 9
    With one "0" in the end: 9
    With two "0": 9
    Without "0": 3 x 9 x 8 = 216
    9 + 9 + 9 + 216 = 243.

12. "1" at ten's digit (no "2" at unit's digit): 9. "1" at unit's digit (no "2"at ten's digit): 8. Removing one duplicate ("11"), 8 + 9 – 1 = 16.

13. 29. Keyword is "or".

14. Three 24-cans, one 12-cans, and one 6-cans. The total number of packs is 3 + 1 + 1 = 5.

15. 6 / (1 – ¾) = 24 chairs. 24 x (¾) / (2/3) = 27 persons.

16. Among five women, there are two single women and three married women. One married woman has one married man. So, the fraction is 3 / (2 + 3 + 3) = 3/8.

17. 1000 = 6 x 166 + 4 ➔ S = 4
    166 = 6 x 27 + 4 ➔ U = 4
    27 = 6 x 4 + 3 ➔ I = 3 and A = 4

18. 8 R ➔ 20 G, 18 G ➔ 30 P, 28 P ➔ 21 Y, with 1 R, 2 G, and 2 P left. There are 21 + 1 + 2 + 2 = 26 tokens in all.

19. Adding two coins to the box, the total number of coins can be equally divided among six or five people. The smallest number of coins is 5 x 6 – 2 = 28, which can be evenly divided among seven people. There will be 0 coin left.

Count with Venn Diagram

1. (2007 – 1001) / 2 + 1001 = 1504.

2. 20 + 10 – 5 = 25 students, in total. Note no student in this class can speak neither English nor Spanish.

3. The number of adults that don't own a motorcycle is 351 – 45 = 306. One who doesn't own a motorcycle owns a car, the answer is 306.

4. 20 + 26 – 39 = 7.

5. Let x be the total number of students in the band. First, we add the number of kids in band, leadership, and track: $x + 15 + 23$. However, this count of the total number of kids double counts the kids in two or more activities, so we must subtract the number of kids in two or more activities: $x + 15 + 23 - 4 - 6 - 3$. Now, for those kids who attend all three activities, we've added them three times (with x, 15, and 23), but we've *also* subtracted them three times (in the 4, 6, and 3). Since these kids also are part of the math team, we must add

them back once: x+15+23-4-6-3+2. This is equal to the total number of kids on the math team, so:

$$40 = x + 15 + 23 - 4 - 6 - 3 + 2$$
$$40 = x + 27$$
$$x = 13$$

Therefore, the total number of people in the band is equal to 13.

6. M − Z − X + A

7. 37 − (6 + 14 + 15 + 15 − 7 − 4 − 5) = 3.

8. There are 100 + 80 − 60 = 120 females in either band or orchestra, so there are 230 − 120 = 110 males in either band or orchestra. 100 + 80 − 110 = 70 males are in both band and orchestra. Thus, the number of males in band but not orchestra is 80 − 70 = 10.

9. 2/5 + ¾ - 1 = 3/20. The minimum number of people in the room wearing both is 3.

Chessboard

1. This is a number series, 1, 4, 7, ..., 1 + 3 x (64 − 1); (1 + (1 + 3 x 63)) x 64 / 2 = 6112.

2. [B]

3. 1 x n (n=1, 2, ..., 8) ➔ 8 x (8 + 1) x 8 / 2
    2 x n (n=1, 2, ..., 8) ➔ 7 x (8 + 1) x 8 / 2
    ......
    8 x n (n=1, 2, ..., 8) ➔ 1 x (8 + 1) x 8 / 2
    1296 including squares

4. 64/2/2/2/2 = 4.
    33 34
    41 42

5. 120.

6. 9.

Coin, Die, and Card

1. (A) (1/16) x (4/52) = 1 / 208.

2. Out of all 8 possible outcomes, only HHT, THH, HHH, three cases are qualifying. Answer is 3/8.

3. $((36 - 6) / 2 + 6) / 36 = 7/12$.

4. $(2 \times C(4, 2) + 4) / C(8, 2) = (12 + 4) / 28 = 4/7$.

5. The outcome is always divisible by 6, so answer is 1.

6. There are a total of $2^4 = 16$ possible outcomes. Probability is: $(6 + 4 + 1) / 16 = 11/16$.

7. $1 - (5/6) \times (5/6) = 11/36$.

8. No head: 1 outcome. One head: 2 outcomes. All: $2 \times 4 = 8$ outcomes. Probability $= 3/8$.

9. $(5 + 30) / 60 = 7/12$.

10. There are 10 configurations out of 64, so probability is $10/64 = 5/32$.

11. $(6 \times 6 - 6) / 2 = 5/12$.

12. $1 - (2 \times 1) / (3 \times 3) = 7/9$.

13. For the sum to be even, the two selected numbers must have the same parity (both even or both odd). The first spinner has 2 odd numbers and 2 even. No matter what the second spinner is, there is a ½ chance the first spinner lands on a number with the same parity, so the probability of an even sum is ½. If we express this in a math expression: $(1/2)(1/3) + (1/2)(2/3) = ½$.

14. $1 \times (5 - 1) / (10 - 1) = 4/9$.

15. $(4 / (9 + 3 + 4 + 1)) \times (3 / (9 + 3 + 3 + 1)) = 3/68 = a/b$. $a + b = 71$.

16. $1 \times 12/51 = 4/17$.

17. Total number of possible outcomes is $6 \times 6 = 36$. To get sum of x, the number of outcomes is 3. The x can be 4 or 10.

18. $x / (7 + x) = y / 12$ ➜ $x = y$ and $7 + x = 12$ ➜ $x = y = 5$, so $xy = 25$.

19. $8 \times 8 \times 7 \times 5 / 9000 = 56/225$

Game of Probability

1. ½.

2. $(1/3) \times (\frac{3}{4}) + (2/3) \times (\frac{1}{4}) = 5/12$.

3. $2/3$.

4. $(1/3) \times (2 \times \frac{1}{4}) + (1 - 1/3) \times (1/4) = 1/6 + 1/6 = 1/3$.

5. $25/91$.

6. Let's look at the expected outcomes:
   (a) When Colin throws 2, Ozzie throws 1, 1 ➜ 1 count
   (b) When Colin throws 3, Ozzie throws 1,2 and 2,1, plus result from (a) ➜ 3 counts
   (c) When Colin throws 4, Ozzie throws 1,3; 3,1; 22; plus result from (b) ➜ 6 counts
   (d) When Colin throws 5, Ozzie throws 1,4; 4,1; 2,3; 3,2; plus result from (c) ➜ 10 counts
   (e) When Colin throws 6, Ozzie throws 1,5; 5,1; 2,4; 4,2; 3,3; plus result from (d) ➜ 15 counts
   $(1 + 3 + 6 + 10 + 15) / (6 \times 6 \times 6) = 35/216$.

Count in Graph

1. 42.

2. There are 61 vertical columns with a length of 32 toothpicks, and there are 33 horizontal rows with a length of 60 toothpicks. An effective way to verify this is to try a small case, i.e. a 2 x 3 grid of toothpicks. Thus, our answer is 61 x 32 + 33 x 60 = 3932.

3. There are 2 ways to place the remaining A, 2 ways to place the remaining B, and 1 way to place the remaining C. So, total is 2 x 2 x 1 = 4.

4. Since the bricks are 1 foot high, there will be 7 rows. To minimize the number of blocks used, rows 1, 3, 5, and 7 will look like the bottom row of the picture, which takes 100 /2 = 50 bricks to construct. Rows 2, 4, and 6 will look like the upper row pictured, which has 49 2-foot bricks in the middle, and two 1-foot bricks on each end for a total of 51 bricks. Four rows of 50 bricks and three rows of 51 bricks totals 4 x 50 + 3 x 51 = 353 bricks.

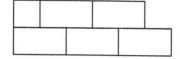

5. There is 1 way to get from C to N. There is only one way to get from D to N, which is DCN. A can only go to C or D, which each only have 1 way to get to N each, there are 1 + 1 = 2 ways to get from A to N. B can only go to A, C or N, and A only has 2 ways to get to N, C only has 1 way and to get from B to N is only 1 way, there are 2 + 1 + 1 = 4 ways to get from B to N. M

108

can only go to either B or A, A has 2 ways and B has 4 ways, so M has 4 + 2 = 6 ways to get to N, see diagram as below.

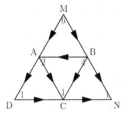

6. The number of ways to get from Samantha's house to City Park is $\binom{3}{1} = 3$

    The number of ways to get from City Park to school is $\binom{4}{2} = 6$
    There is only one way to go through City Park, so the number of different ways to go from Samantha's house to City Park to school 3 x 6 = 18.

Logical Reasoning

1. A = 20/40, B = 20 / 100; C = 18 / 100. A + B + C = 22/25.

2. Every month can have up to 31 days. The smallest two digit primes are 11, 13, and 17. There are a few different sums that can be deduced from the following numbers, which are 24, 30, and 28, all of which represent the three days (less than 31).
    Since Brittany says the other two people's uniform numbers is earlier, that means Caitlin and Ashley's numbers must add up to 24.
    Caitlin says the other two people's uniform numbers is later, so the sum must add up to 30. This leaves 28 as today's date. Since Caitlin was referring to the uniform wearers 13 and 17, the number she wear is 11.

3. The only way to get Ben's score is with 1 + 3 = 4. Cindy's score can be made of 3 + 4 or 2 + 5, but since Ben already hit the 3, Cindy hit 2 + 5 = 7. Similar, Dave's darts were in the region 4 + 7 = 11. Lastly, because there is no 7 left, Alice must have hit the regions 6 + 10 = 16 and Ellen 8 + 9 = 17.

4. According to rule 1, the largest number 12, can be second or third. According to rule 2, because there are five places, the smallest number 2 can be either third or fourth. The median, 6 can be second, third, or fourth. Because we know the middle three numbers, the first and last numbers are 4 and 9, disregarding their order. Their average is (4 + 9) / 2 = 6.5.

5. Jim has brown eyes and blonde hair. If you look for anybody who has brown eyes **or** blonde hair, you find that Nadeen, Austin, and Sue are Jim's possible siblings. However, the children in the same family have at least one common characteristics. Since Austin and Sue both have blonde hair, Nadeen is ruled out. Therefore Austin and Sue are Jim's siblings.

6. Since the only other score Quay knows is Kaleana's, and he knows that two of them have the same score, Quay and Kaleana must have the same score. Use K and Q represent their score respectively, then K = Q.
    Marty knows that he didn't get the lowest score, and the only other score he knows is Kaleana's, so Marty must know that Kaleana has a lower score than his. Use M and K represent their score respectively, then M > K.
    Shana knows that she didn't get the highest score, and the only other score she knows is Kaleana's, so Shana must know that Kaleana has a higher score than her. Use S and K represent their score respectively, then S < K.
    Putting things together, we now know S < Q < M.

7. First, note that P beats Q, R, T, and S, by transitivity. ➜ P is the first, and not the 3rd.
    Similarly, S is beaten by P, Q, and T, by transitivity. ➜ S is the fourth or fifth, not in third. All of the others could be in third, as the following sequences show: PTQRS, PTRQS, PRTQS. Therefore, the answer is P and S.

8. We know that Carl does not sit next to Bret, so he must sit in seat #1. Since Abby is not between Bret and Carl, she must sit in seat #4. Finally, Dana has to take the last seat available, which is #2.

9. First, let's assume that Alan gets an A. From Alan's statement, Beth would also get an A. But from Beth's statement, Carlos would get an A. And from Carlos's statement, Diana would also get an A. So all 4 would get A's, but the problem said only 2 got A's.
    Secondly, let's assume that Beth gets an A. From her statement, we know that Carlos get an A, and from his statement we know that Diana gets an A. But that makes 3, which is not 2.
    If Carlos gets an A, then Diana gets an A. That makes 2, so the right answer is Carlos and Diana. Note that although Beth said "If I get an A, then Carlos will get an A.", that does NOT mean that "If Carlos gets an A, then I will get an A."

Number Base

1. 3727_8

2. 11 digits. $2^{10} < 2013_{10} < 2^{11}$.

3. 5/8. All samples: 8 through 15 in ten base; Expected samples: 11 through 15 in ten base.

Algebraic Way of Thinking

1. $x + y + z = 12$, $x + 3y + 4z = 24$, integer: x,y,z > 1 ➜ y=3, z=2, x=7 ➜ 7 pairs of $1 socks.

2. $x / (20/60) + 18 = x / (12/60)$ ➜ x = 9 miles.

3. 995 dollars.

4. $1/10 - 1/x = 1/20 + 1/x$ ➜ $x = 40$ minutes.

5. 300D/P.

6. 27.

7. 12.

8. 13, 14.

9. $bg + gb + 42 = (b + g)(b + g + 1)$ ➜ $b + g = 8$.

10. $(135 - x)r = 810$, $(135 - 2x)r = 135 + 270$ ➜ $x = 45$ kg.

11. 40.

12. $40 <= (12 - x) \times 6 + 12$, AND $6 \times (12 - x) < 40$ AND $5 \times (12 - x) + 12 < 40$ ➜ $x = 7$ ➜ sum of all possible x = 7.

13. 49Pi.

14. 9.

15. $5N + 10D + 25Q = 5Q + 10N + 25D$, $Q + D + N = 10$ ➜ Q=4, D=5, N=1 ➜ 155 cents.

16. 8.

17. $n = 9r + q$ ➜ $n = 8r + (r + q) = 8r + 13$. Given $0 < q < 9$ and $r >= 5$, the minimum of $n = 8 \times 5 + 13 = 53$.

18. Define total number of gold coins is gc, then $(gc + 18) / 9 = (gc - 3) / 6$ ➜ gc = 45.

Newton's Cow and Grass

1. First we find the amount of grass that grows each day:
   $(17 \times 30 - 19 \times 24) \div (30 - 24) = 9$ cow's grass per day.
   Secondly, we find the initial amount of grass before any cow eats:
   $(17 \times 30 - 9 \times 30) = 240$. This is an amount of grass that 240 cows could eat in a day.

111

We let the initial number of cows before 4 of them are sold be *x*.
Then $(x-9) \times 6 + (x-4-9) \times 2 = 240$.
x = 40 cows initially.

Number and Words

- 3 Sides on a Triangle
- 4 Seasons in a Year
- 5 Toes on each Foot
- 6 Sides on a Cube
- 7 Days in a Week
- 8 Sides on a Stop Sign
- 9 Planets in the Solar System
- 12 Edges on a Cube
- 13 Cards in a Suit
- 18 Holes in a Golf Course
- 24 Hours in a Day
- 26 Letters of the Alphabet
- 29 Days in February in a Lunar Year
- 52 Cards in a Deck (without the Jokers)
- 54 Cards in a Deck (with the Jokers)
- 60 Seconds in a Minutes
- 64 Squares on a Chess Board
- 88 Piano Keys
- 90 Degrees in a Right Angle
- 206 Bones in the Body
- 360 Degrees in a Circle
- 747 Boeing Airplane
- 1001 Arabian Nights
- 1040 Tax Return Form

CPSIA information can be obtained
at www.ICGtesting.com
Printed in the USA
LVOW04s0341161217
559964LV00007B/386/P